Princip
Chemi

A GUIDE

IUPAC RI

Principles of Chemical Nomenclature
A GUIDE TO IUPAC RECOMMENDATIONS

G.J. LEIGH OBE
The School of Chemistry, Physics
and Environmental Science,
University of Sussex, Brighton, UK

H.A. FAVRE
Université de Montréal
Montréal, Canada

W.V. METANOMSKI
Chemical Abstracts Service
Columbus, Ohio, USA

Edited by G.J. Leigh

**Blackwell
Science**

© 1998 by
Blackwell Science Ltd
Editorial Offices:
Osney Mead, Oxford OX2 0EL
25 John Street, London WC1N 2BL
23 Ainslie Place, Edinburgh EH3 6AJ
350 Main Street, Malden
    MA 02148-5018, USA
54 University Street, Carlton
    Victoria 3053, Australia
10, Rue Casmir Delavigne
    75006 Paris, France

Other Editorial Offices:
Blackwell Wissenschafts-Verlag GmbH
Kurfürstendamm 57
10707 Berlin, Germany

Blackwell Science KK
MG Kodenmacho Building
7–10 Kodenmacho Nihombashi
Chuo-ku, Tokyo 104, Japan

First published 1998

Set by Semantic Graphics, Singapore
Printed and bound in Great Britain
by MPG Books Ltd, Bodmin, Cornwall

The Blackwell Science logo is a
trade mark of Blackwell Science Ltd,
registered at the United Kingdom
Trade Marks Registry

DISTRIBUTORS

Marston Book Services Ltd
PO Box 269
Abingdon
Oxon OX14 4YN
(*Orders*: Tel: 01235 465500
          Fax: 01235 465555)

USA
Blackwell Science, Inc.
Commerce Place
350 Main Street
Malden, MA 02148-5018
(*Orders*: Tel: 800 759 6102
           781 388 8250
      Fax: 781 388 8255)

Canada
Copp Clark Professional
200 Adelaide St West, 3rd Floor
Toronto, Ontario M5H 1W7
(*Orders*: Tel: 416 597-1616
           800 815-9417
      Fax: 416 597-1617)

Australia
Blackwell Science Pty Ltd
54 University Street
Carlton, Victoria 3053
(*Orders*: Tel: 3 9347 0300
      Fax: 3 9347 5001)

A catalogue record for this title
is available from the British Library

ISBN 0-86542-685-6

Library of Congress
Cataloging-in-publication Data

Leigh, G. J.
    Principles of chemical nomenclature : a guide to
IUPAC recommendations / G.J. Leigh,
H.A. Favre, W.V. Metanomski.
        p.   cm.
    Includes bibliographical references
    and index.
    ISBN 0-86542-685-6
    1. Chemistry—Nomenclature.
    I. Favre, H.A.   II. Metanomski, W.V.
III. International Union of Pure and Applied
Chemistry.   IV. Title.
QD7.L44   1997
540′.14—dc21   97-28587
                                        CIP

Contents

CONTENTS

Preface

This book arose out of the convictions that IUPAC nomenclature needs to be made as accessible as possible to teachers and students alike, and that there is an absence of relatively complete accounts of the IUPAC 'colour' books suited to school and undergraduate audiences. This is not to decry in any way the efforts of organisations such as the Association for Science Education (ASE) in the UK, but what we wished to produce was a version of IUPAC rules that would be relatively complete and allow the beginner to explore and learn about nomenclature as much or as little as desired.

Initially, it was intended to produce a book that would cover all IUPAC colour books and encompass much more than what is conventionally regarded as nomenclature, e.g. dealing also with units, kinetics and analysis. A committee consisting of C. J. H. Schutte (South Africa), J. R. Bradley (South Africa), T. Cvitaš (Croatia), S. Głąb (Poland), H. A. Favre (Canada) and G. J. Leigh (UK) was set up to produce a draft of this book. Later, they were joined by W. V. Metanomski (USA). When the first draft had been prepared, it was evident that the conventional nomenclature section was so large that it unbalanced the whole production.

Finally, it was decided to prepare two texts, one following the original proposal, but with a much reduced nomenclature content in order to restore the balance, and a second, this volume, that would attempt to survey the current IUPAC nomenclature recommendations in organic, inorganic and macromolecular chemistry and also include some basic biochemical nomenclature. This was undertaken by Favre, Leigh and Metanomski, with the final editing being undertaken by Leigh.

It is hoped that this volume will more than cover all the nomenclature requirements of students at pre-University and early undergraduate levels in most countries. It should also enable University students and teachers to learn the basic principles of nomenclature methods so that they can apply them accurately and with confidence. It will probably be too advanced for school students, but should be useful for their teachers.

Specialists in nomenclature recognise two different categories of nomenclature. Names that are arbitrary (including the names of the elements, such as sodium and hydrogen) as well as laboratory shorthand names (such as diphos and LithAl) are termed trivial names. This is not a pejorative or dismissive term. Trivial nomenclature contrasts with systematic nomenclature, which is developed according to a set of prescribed rules. However, nomenclature, like any living language, is growing and changing. This is reflected by the fact that IUPAC does not prescribe a single name for each and every compound.

There are several extant systems of nomenclature and many trivial names are still in use. This means that the chemist often has a selection of names from which to choose. IUPAC may prefer some names and allow others, and the name selected should generally be, within reason, a systematic one. Because IUPAC cannot legislate, but can only advise, chemists should feel free to back their own judgement. For example, the systematic name for NH_3 is azane, but it is not recommended for general use in place of the usual 'ammonia'. On the other hand, there seems to be no good reason why chemists generally should not adopt the more systematic phos-phane, rather than phosphine, for PH_3.

Students may find this matter of choice confusing on occasion, which will be a pity. However, there are certain long-established principles that endure, and we hope to have encompassed them in this book.

G. J. Leigh
University of Sussex
June 1997

1 Introduction

Chemical nomenclature is at least as old as the pseudoscience of alchemy, which was able to recognise a limited number of reproducible materials. These were assigned names that often conveyed something of the nature of the material (vitriol, oil of vitriol, butter of lead, aqua fortis . . .). As chemistry became a real science, and principles of the modern atomic theory and chemical combination and constitution were developed, such names no longer sufficed and the possibility of developing systematic nomenclatures was recognised. The names of Guyton de Morveau, Lavoisier, Berthollet, Fourcroy and Berzelius are among those notable for early contributions. The growth of organic chemistry in the nineteenth century was associated with the development of more systematic nomenclatures, and chemists such as Liebig, Dumas and Werner are associated with these innovations.

The systematisation of organic chemistry in the nineteenth century led to the early recognition that a systematic and internationally acceptable system of organic nomenclature was necessary. In 1892, the leading organic chemists of the day gathered in Geneva to establish just such a system. The Geneva Convention that they drew up was only partly successful. However, it was the forerunner of the current activities of the International Union of Pure and Applied Chemistry (IUPAC) and its Commission on Nomenclature of Organic Chemistry (CNOC), which has the remit to study all aspects of the nomenclature of organic substances, to recommend the most desirable practices, systematising trivial (i.e. non-systematic) methods, and to propose desirable practices to meet specific problems. The Commission on the Nomenclature of Inorganic Chemistry (CNIC) was established rather later, because of the later systematisation of this branch of the subject, and it now fulfils functions similar to those of CNOC but in inorganic chemistry. In areas of joint interest, such as organometallic chemistry, CNIC and CNOC collaborate. The recommendations outlined here are derived from those of these IUPAC Commissions, and of the Commission on Macromolecular Nomenclature (COMN) and of the International Union of Biochemistry and Molecular Biology (IUBMB).

The systematic naming of substances and presentation of formulae involve the construction of names and formulae from units that are manipulated in accordance with defined procedures in order to provide information on composition and structure. There are a number of accepted systems for this, of which the principal ones will be discussed below. Whatever the pattern of nomenclature, names and formulae are constructed from units that fall into the following classes:
- Element names, element name roots, element symbols.
- Parent hydride names.
- Numerical prefixes (placed before a name, but joined to it by a hyphen), infixes (inserted into a name, usually between hyphens) and suffixes (placed after a name).
- Locants, which may be letters or numerals, and may be prefixes, infixes or suffixes.
- Prefixes indicating atoms or groups — either substituents or ligands.
- Suffixes in the form of a set of letters or characters indicating charge.
- Suffixes in the form of a set of letters indicating characteristic groups.
- Infixes in the form of a set of letters or characters, with various uses.

• Additive prefixes: a set of letters or characters indicating the formal addition of particular atoms or groups to a parent molecule.

• Subtractive suffixes and/or prefixes: a set of letters or characters indicating the absence of particular atoms or groups from a parent molecule.

• Descriptors (structural, geometric, stereochemical, etc.).

• Punctuation marks.

The uses of all these will be exemplified in the discussion below.

The material discussed here is based primarily on *A Guide to IUPAC Nomenclature of Organic Chemistry, Recommendations 1993*, issued by CNOC, on the *Nomenclature of Inorganic Chemistry, Recommendations 1990* (the Red Book), issued by CNIC, on the *Compendium of Macromolecular Chemistry* (the Purple Book), issued in 1991 by COMN, and on *Biochemical Nomenclature and Related Documents*, 2nd Edition 1992 (the White Book), issued by IUBMB.

In many cases, it will be noted that more than one name is suggested for a particular compound. Often a preferred name will be designated, but as there are several systematic or semi-systematic nomenclature systems it may not be possible, or even advisable, to recommend a unique name. In addition, many non-systematic (trivial) names are still in general use. Although it is hoped that these will gradually disappear from the literature, many are still retained for present use, although often in restricted circumstances. These restrictions are described in the text. The user of nomenclature should adopt the name most suitable for the purpose in hand.

2 Definitions

An element (or an elementary substance) is matter, the atoms of which are alike in having the same positive charge on the nucleus (or atomic number).

In certain languages, a clear distinction is made between the terms 'element' and 'elementary substance'. In English, it is not customary to make such nice distinctions, and the word 'atom' is sometimes also used interchangeably with element or elementary substance. Particular care should be exercised in the use and comprehension of these terms.

An atom is the smallest unit quantity of an element that is capable of existence, whether alone or in chemical combination with other atoms of the same or other elements.

The elements are given names, of which some have origins deep in the past and others are relatively modern. The names are trivial. The symbols consist of one, two or three roman letters, often but not always related to the name in English.

Examples
1. Hydrogen   H
2. Argon   Ar
3. Potassium   K
4. Sodium   Na
5. Chlorine   Cl
6. Ununquadium   Uuq

For a longer list, see Table 2.1. For the heavier elements as yet unnamed or unsynthesised, the three-letter symbols, such as Uuq, and their associated names are provisional. They are provided for temporary use until such time as a consensus is reached in the chemical community that these elements have indeed been synthesised, and a trivial name and symbol have been assigned after the prescribed IUPAC procedures have taken place.

When the elements are suitably arranged in order of their atomic numbers, a Periodic Table is generated. There are many variants, and an IUPAC version is shown in Table 2.2.

An atomic symbol can have up to four modifiers to convey further information. This is shown for a hypothetical atomic symbol X:

$$^D_C X ^A_B$$

Modifier A indicates a charge number, which may be positive or negative (when element X is more properly called an ion). In the absence of modifier A, the charge is assumed to be zero. Alternatively or additionally, it can indicate the number of unpaired electrons, in which case the modifier is a combination of an arabic numeral and a dot. The number 'one' is not represented.

Table 2.1 Names, symbols and atomic numbers of the atoms (elements).

Name	Symbol	Atomic number	Name	Symbol	Atomic number
Actinium	Ac	89	Mercury[6]	Hg	80
Aluminium	Al	13	Molybdenum	Mo	42
Americium	Am	95	Neodymium	Nd	60
Antimony[1]	Sb	51	Neon	Ne	10
Argon	Ar	18	Neptunium	Np	93
Arsenic	As	33	Nickel	Ni	28
Astatine	At	85	Niobium	Nb	41
Barium	Ba	56	Nitrogen[7]	N	7
Berkelium	Bk	97	Nobelium	No	102
Beryllium	Be	4	Osmium	Os	76
Bismuth	Bi	83	Oxygen	O	8
Bohrium	Bh	107	Palladium	Pd	46
Boron	B	5	Phosphorus	P	15
Bromine	Br	35	Platinum	Pt	78
Cadmium	Cd	48	Plutonium	Pu	94
Caesium	Cs	55	Polonium	Po	84
Calcium	Ca	20	Potassium[8]	K	19
Californium	Cf	98	Praseodymium	Pr	59
Carbon	C	6	Promethium	Pm	61
Cerium	Ce	58	Protactinium	Pa	91
Chlorine	Cl	17	Radium	Ra	88
Chromium	Cr	24	Radon	Rn	86
Cobalt	Co	27	Rhenium	Re	75
Copper[2]	Cu	29	Rhodium	Rh	45
Curium	Cm	96	Rubidium	Rb	37
Dubnium	Db	105	Ruthenium	Ru	44
Dysprosium	Dy	66	Rutherfordium	Rf	104
Einsteinium	Es	99	Samarium	Sm	62
Erbium	Er	68	Scandium	Sc	21
Europium	Eu	63	Seaborgium	Sg	106
Fermium	Fm	100	Selenium	Se	34
Fluorine	F	9	Silicon	Si	14
Francium	Fr	87	Silver[9]	Ag	47
Gadolinium	Gd	64	Sodium[10]	Na	11
Gallium	Ga	31	Strontium	Sr	38
Germanium	Ge	32	Sulfur[11]	S	16
Gold[3]	Au	79	Tantalum	Ta	73
Hafnium	Hf	72	Technetium	Tc	43
Hassium	Hs	108	Tellurium	Te	52
Helium	He	2	Terbium	Tb	65
Holmium	Ho	67	Thallium	Tl	81
Hydrogen[4]	H	1	Thorium	Th	90
Indium	In	49	Thulium	Tm	69
Iodine	I	53	Tin[12]	Sn	50
Iridium	Ir	77	Titanium	Ti	22
Iron[5]	Fe	26	Tungsten[13]	W	74
Krypton	Kr	36	Ununbiium	Uub	112
Lanthanum	La	57	Ununhexium	Uuh	116
Lawrencium	Lr	103	Ununnilium	Uun	110
Lead	Pb	82	Ununoctium	Uuo	118
Lithium	Li	3	Ununpentium	Uup	115
Lutetium	Lu	71	Ununquadium	Uuq	114
Magnesium	Mg	12	Ununseptium	Uus	117
Manganese	Mn	25	Ununtriium	Uut	113
Meitnerium	Mt	109	Unununium	Unu	111
Mendelevium	Md	101	Uranium	U	92

Continued.

Table 2.1 (*Continued.*)

Name	Symbol	Atomic number	Name	Symbol	Atomic number
Vanadium	V	23	Yttrium	Y	39
Xenon	Xe	54	Zinc	Zn	30
Ytterbium	Yb	70	Zirconium	Zr	40

[1] Symbol derived from the Latin name stibium.
[2] Symbol derived from the Latin name cuprum.
[3] Symbol derived from the Latin name aurum.
[4] The hydrogen isotopes ^2H and ^3H are named deuterium and tritium, respectively, for which the symbols D and T may be used.
[5] Symbol derived from the Latin name ferrum.
[6] Symbol derived from the Latin name hydrargyrum.
[7] The name azote is used to develop names for some nitrogen compounds.
[8] Symbol derived from the Latin name kalium.
[9] Symbol derived from the Latin name argentum.
[10] Symbol derived from the Latin name natrium.
[11] The Greek name theion provides the root 'thi' used in names of sulfur compounds.
[12] Symbol derived from the Latin name stannum.
[13] Symbol derived from the Germanic name wolfram.

Examples
7. Na^+ 10. Cl^-
8. Ca^{2+} 11. O^{2-}
9. N^{3-} 12. $N^{.2-}$

Modifier B indicates the number of atoms bound together in a single chemical entity or species. If B is 1, it is not represented. In an empirical formula (see below) it can be used to indicate relative proportions.

Examples
13. P_4
14. Cl_2
15. S_8
16. C_{60}

Modifier C is used to denote the atomic number, but this space is generally left empty because the atomic symbol necessarily implies the atomic number.

Modifier D is used to show the mass number of the atom being considered, this being the total number of neutrons and protons considered to be present in the nucleus. The number of protons defines the element, but the number of neutrons in atoms of a given element may vary. Any atomic species defined by specific values of atomic number and mass number is termed a nuclide. Atoms of the same element but with different atomic masses are termed isotopes, and the mass number can be used to designate specific isotopes.

Examples
17. ^{31}P
18. ^1H, ^2H (or D), ^3H (or T)
19. ^{12}C

Table 2.2 IUPAC Periodic Table of the Elements.

1	2	3	4	5	6	7	8	9	10	11	12	13	14	15	16	17	18	n
1 H																	2 He	1
3 Li	4 Be											5 B	6 C	7 N	8 O	9 F	10 Ne	2
11 Na	12 Mg											13 Al	14 Si	15 P	16 S	17 Cl	18 Ar	3
19 K	20 Ca	21 Sc	22 Ti	23 V	24 Cr	25 Mn	26 Fe	27 Co	28 Ni	29 Cu	30 Zn	31 Ga	32 Ge	33 As	34 Se	35 Br	36 Kr	4
37 Rb	38 Sr	39 Y	40 Zr	41 Nb	42 Mo	43 Tc	44 Ru	45 Rh	46 Pd	47 Ag	48 Cd	49 In	50 Sn	51 Sb	52 Te	53 I	54 Xe	5
55 Cs	56 Ba	57–71 La–Lu	72 Hf	73 Ta	74 W	75 Re	76 Os	77 Ir	78 Pt	79 Au	80 Hg	81 Tl	82 Pb	83 Bi	84 Po	85 At	86 Rn	6
87 Fr	88 Ra	89–103 Ac–Lr	104 Rf	105 Db	106 Sg	107 Bh	108 Hs	109 Mt	110 Uun	111 Uuu	112 Uub	113 Uut	114 Uuq	115 Uup	116 Uuh	117 Uus	118 Uuo	7

																n
57 La	58 Ce	59 Pr	60 Nd	61 Pm	62 Sm	63 Eu	64 Gd	65 Tb	66 Dy	67 Ho	68 Er	69 Tm	70 Yb	71 Lu		6
89 Ac	90 Th	91 Pa	92 U	93 Np	94 Pu	95 Am	96 Cm	97 Bk	98 Cf	99 Es	100 Fm	101 Md	102 No	103 Lr		7

Note that of all the isotopes of all the elements, only those of hydrogen, ^2H and ^3H, also have specific atomic symbols, D and T, with associated names deuterium and tritium.

Elements fall into various classes, as laid out in the Periodic Table (Table 2.2). Among the generally recognised classes are the Main Group elements (Groups 1, 2, 13, 14, 15, 16, 17 and 18), the two elements of lowest atomic number in each group being designated typical elements. The elements of Groups 3–11 are transition elements. The first element, hydrogen, is anomalous and forms a class of its own. Other more trivial designations (alkali metals, halogens, etc.) are recognised, but these names are not often used in nomenclature. For more information, consult an appropriate textbook.

Only a few elements form a monoatomic elementary substance. The majority form polyatomic materials, ranging from diatomic substances, such as H_2, N_2 and O_2, through polyatomic species, such as P_4 and S_8, to infinite polymers, such as the metals. These polyatomic species, where the degree of aggregation can be precisely defined, are more correctly termed molecules. However, the use of the term 'element' is not restricted to the consideration of elementary substances. Compounds are composed of atoms of the same or of more than one kind of element in some form of chemical combination. Thus water is a compound of the elements hydrogen and oxygen. The molecule of water is composed of three atoms, two of which are of the element hydrogen and one of the element oxygen. It should be noted here, again, that the term 'element' is one that is sometimes considered to be an abstraction. It implies the essential nature of an atom, which is retained however the atom may be combined, or in whatever form it exists. An elementary substance is a physical form of that element, as it may be prepared and studied.

Molecules can also be charged. This is not common in elementary substances, but where some molecules or atoms are positively charged (these as a class are called 'cations') they must be accompanied by negative molecules or atoms (anions) to maintain electroneutrality.

Many elements can give rise to more than one elementary substance. These may be substances containing assemblages of the same mono- or poly-atomic unit but arranged differently in the solid state (as with tin), or they may be assemblages of different polyatomic units (as with carbon, which forms diamond, graphite and the fullerenes, and with sulfur and oxygen). These different forms of the element are referred to as allotropes. Their common nomenclature is essentially trivial, but attempts have been made to develop systematic nomenclatures, especially for crystalline materials. These attempts are not wholly satisfactory.

Throughout this discussion, we have been considering pure substances, i.e. substances composed of a single material, whether element or compound. A compound may be molecular or ionic, or both. A compound is a single chemical substance. To anticipate slightly, sodium chloride is an ionic compound that contains two atomic species, Na^+ and Cl^-. If a sample of sodium chloride is formally manipulated to remove some Cl^- ions and replace them by Br^- ions in equivalent number, the resultant material is a mixture. The same is true of a sample containing neutral species such as P_4, S_8 and C_6H_6.

Pure substances (be they elementary or compound) and mixtures are usually solids, liquids or gases, and they may even take some rarer form. These forms are termed states of matter and are not strictly the province of nomenclature. However, to indicate by a name or a formula whether a substance is a solid, liquid or gas, the letters s, g or l are used. For more details, see the Green Book (*Quantities, Units and Symbols in Physical Chemistry*, 2nd Edition, Blackwell Scientific Publications, Oxford, 1993).

Examples
20. $H_2O(l)$
21. $H_2O(g)$
22. $H_2O(s)$

3 Formulae

INTRODUCTION

The basic materials of systematic chemical nomenclature are the element names and symbols, which are, of themselves, trivial, with the exception of the systematic, provisional names and symbols for the elements of atomic number greater than 109. These provisional names will be superseded eventually by trivial names and symbols. In any case, they make little impact on general chemical practice.

The simplest way to represent chemical substances is to use formulae, which are assemblages of chemical symbols. Formulae are particularly useful for listing and indexing and also when names become very complex. The precise form of a formula selected depends upon the use to which it is to be put.

EMPIRICAL FORMULAE

The simplest kind of formula is a compositional formula or empirical formula, which lists the constituent elements in the atomic proportions in which they are present in the compound. For such a formula to be useful in lists or indexes, an order of citation of symbols (hierarchy) must be agreed. Such hierarchies, often designated seniorities or priorities, are commonly used in nomenclature. For lists and indexes, the order is now generally recommended to be the alphabetical order of symbols, with one very important exception. Because carbon and hydrogen are always present in organic compounds, C is always cited first, H second and then the rest, in alphabetical order. In non-carbon-containing compounds, strict alphabetical order is adhered to.

Note that molecular or ionic masses cannot be calculated from empirical formulae.

Examples
1. ClK
2. CaO_4S
3. $C_6FeK_4N_6$
4. NS
5. $C_{10}H_{10}ClFe$
6. CH_2
7. CHO

MOLECULAR FORMULAE

Molecular formulae for compounds consisting of discrete molecules are formulae according with the relative molar mass or relative molecular mass or molecular weight for the structure.

Examples
1. N_4S_4
2. S_2Cl_2
3. C_2H_6

Polyatomic ions are treated similarly, although the charge must also be indicated. These formulae tell nothing about structure. As soon as structural information is combined with the formula, these simple rules need to be amplified.

It should be noted that the discussion so far has assumed that all compounds are stoichiometric, i.e. that all the atomic or molecular proportions are integral. It has become increasingly clear that many compounds are to some degree non-stoichiometric. These rules fail for non-stoichiometric compounds, for which further formalisms need to be developed. Electroneutrality must, of course, be maintained overall in such compounds, in one way or another. For example, in an ionic compound where there is apparently a deficit of negative ions, the consequent formal excess of cations may be neutralised by the presence of an appropriate number of cations of the form $M^{(n-1)+}$ rather than of the prevalent form M^{n+}. Various stratagems have been used to represent this kind of situation in formulae, although not yet in names. For details, the reader is referred to the *Nomenclature of Inorganic Chemistry*, Chapter 6.

Examples
1. $\sim FeS$
2. $Co_{1-x}O$
3. $(Li_2, Mg)Cl_2$
4. $Fe_{1.05}Li_{3.65}Ti_{1.30}O_6$

3.4 STRUCTURAL FORMULAE

Structural formulae give information about the way atoms in a molecule or ion are connected and arranged in space.

Examples

1. 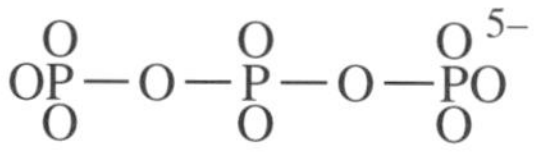
$$O\overset{O}{\underset{O}{P}}-O-\overset{O}{\underset{O}{P}}-O-\overset{O}{\underset{O}{P}}O^{5-} \quad \text{or} \quad \left(O\overset{O}{\underset{O}{P}}-O-\overset{O}{\underset{O}{P}}-O-\overset{O}{\underset{O}{P}}O\right)^{5-}$$

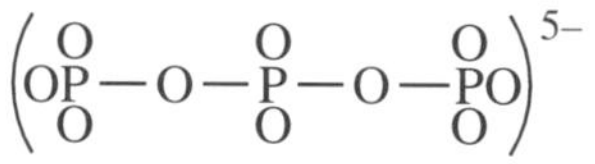

2. 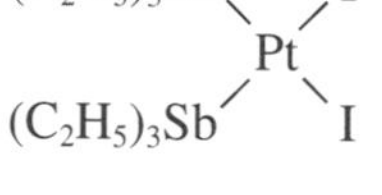
$$\begin{array}{c}(C_2H_5)_3Sb\diagdown \qquad \diagup I \\ Pt \\ (C_2H_5)_3Sb\diagup \qquad \diagdown I\end{array}$$

Attempts may be made to represent the structure in three dimensions.

Example

3. 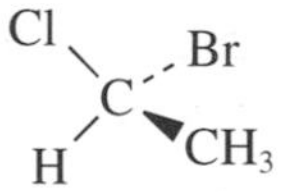
$$\begin{array}{c}Cl\diagdown \quad \diagup Br \\ C \\ H\diagup \quad \blacktriangleright CH_3\end{array}$$

In this example, the full lines represent bonds in the plane of the paper, the dotted line represents a bond pointing below the plane of the paper and the triangular bond points towards the reader. This kind of representation will be discussed in more detail in Section 3.8, p. 21.

In organic chemistry, structural formulae are frequently presented as condensed formulae. This abbreviated presentation is especially useful for large molecules. Another way of presenting structural formulae is by using bonds only, with the understanding that carbon and hydrogen atoms are never explicitly shown.

Examples

4. 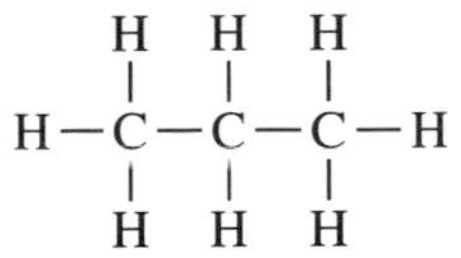
$$H-\underset{\underset{H}{|}}{\overset{\overset{H}{|}}{C}}-\underset{\underset{H}{|}}{\overset{\overset{H}{|}}{C}}-\underset{\underset{H}{|}}{\overset{\overset{H}{|}}{C}}-H \quad \text{or} \quad CH_3\text{-}CH_2\text{-}CH_3 \quad \text{or}$$

5. 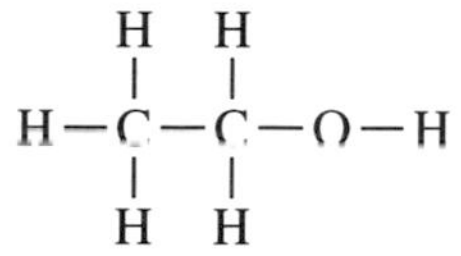
$$H-\underset{\underset{H}{|}}{\overset{\overset{H}{|}}{C}}-\underset{\underset{H}{|}}{\overset{\overset{H}{|}}{C}}-O-H \quad \text{or} \quad CH_3\text{-}CH_2\text{-}OH \quad \text{or}$$

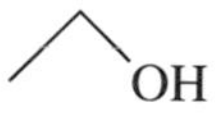

6. $CH_3\text{-}CH_2\text{-}CH_2\text{-}CH_2\text{-}CH_3$ or

7. 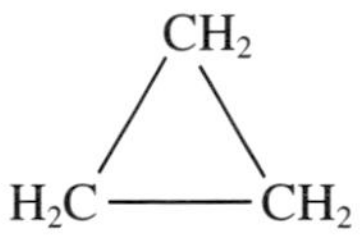
or

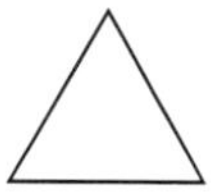

8. 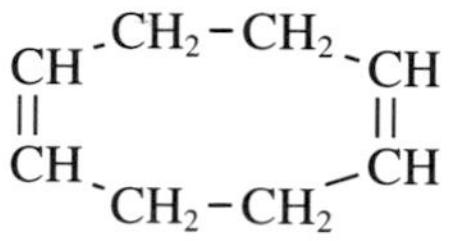
or

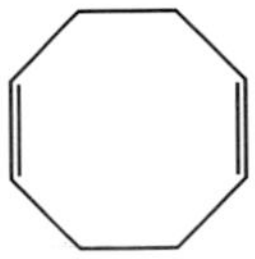

As will be evident from the above examples, and by extrapolation from the rules elicited for species derived from one type of atom, the numbers of groups of atoms in a unit and the charge on a unit are indicated by modifiers in the form of subscripts and superscripts.

Examples

9. $C(CH_3)_4$
10. $CH_3\text{-}[CH_2]_5\text{-}CH_3$
11. $CaCl_2$
12. $[\{Fe(CO)_3\}_3(CO)_2]^{2-}$

Note the use of enclosing marks: parentheses ( ), square brackets [ ] and braces { }. They are used to avoid ambiguity. In the specific case of coordination compounds, square brackets denote a 'coordination entity' (see below). In the organic examples above, the use of square brackets to indicate an unbranched chain is shown. In organic nomenclature generally and in inorganic names, only two classes of enclosing mark are used, ( ) and [ ], with the parentheses being the junior set.

3.5 SEQUENCE OF CITATION OF SYMBOLS

We have already stated that the sequence of atomic symbols in an empirical or molecular formula is arbitrary, but that in the absence of any other requirements a modified alphabetical sequence is recommended. This is primarily a sequence for use in indexes, such as in a book that lists compounds cited by formula.

Where there are no overriding requirements, the following criteria may be adopted for general use. In a formula, the order of citation of symbols is based upon relative electronegativities. Although there is no general confusion about which of, say, Na and Cl represents the more electronegative element, there is no universal scale of electronegativity that is appropriate for all purposes. However, for ionic compounds, cations are always cited before anions. In general, the choice is not so easy. Therefore, the Commission on the Nomenclature of Inorganic Chemistry has recommended the use of Table IV of the *Nomenclature of Inorganic Chemistry* (Table 3.1 of this book) to represent such a scale for nomenclature purposes. The order of citation proposed in a binary compound is from the least electronegative (i.e. most electropositive) to the most electronegative, and the least electronegative element is that encountered last on proceeding through Table 3.1 in the direction of the arrows. Those elements before Al are regarded as electronegative, and those after B as electropositive.

If a formula contains more than one element of each class, the order of citation within each class is alphabetical. Note, however, that 'acid hydrogen' is always regarded as an electropositive element, and immediately precedes the anionic constituents in the formulae of acids.

Examples

1. KCl
2. $Na_2B_4O_7$
3. $IBrCl_2$
4. O_2ClF_3
5. $NaHSO_4$

Where it is known that certain atoms in a molecular ion are bound together to form a group, as with S and O in SO_4^{2-}, these elements can be so grouped in the formula, with or without enclosing marks, depending upon the compound and upon the users' requirements.

Examples

6. HBr
7. H_2SO_4
8. $[Cr(H_2O)_6]Cl_3$
9. $H[AuCl_4]$

Table 3.1 Element sequence.

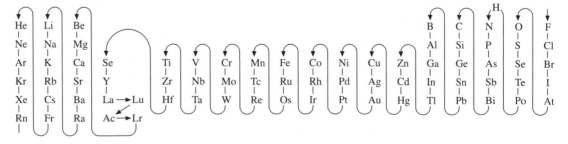

There are various subrules: for example, a single-letter symbol (B) always precedes a two-letter symbol (Be); NH_4 is treated as a two-letter symbol and is listed after Ne. The written alphabetical ordering of a polyatomic group is determined by the first symbol cited: SO_4^{2-} by S; $[Zn(H_2O)_6]^{2+}$ by Zn; NO_3^- by N, etc. A more detailed discussion is given in the *Nomenclature of Inorganic Chemistry*, Chapter 4.

For binary compounds between non-metals (i.e. between elements that are considered to be electronegative), a modified electronegativity sequence (cf. Table 3.1) is adopted, and the least electronegative element is cited first. The sequence of increasing electronegativity is:

Rn Xe Kr Ar Ne He B Si C Sb As P N H Te Se S At I Br Cl O F

For intermetallic compounds, where all the elements can be considered to be electropositive, strict alphabetical ordering of symbols is recommended.

Examples
10. Au_2Bi
11. NiSn

3.6   FORMULAE OF GROUPS

We have already mentioned the formulae for groups, such as SO_4^{2-}, without discussing the principles by which such formulae are assembled. These may (or may not) involve some reference to structure. The general approach is to select one or more atom(s) as the central or characteristic atom(s). This is so whether the ion or group is a coordination entity or not. Thus, I in ICl_4^-, V in VO_2^+ and Si and W in $[SiW_{12}O_{40}]^{4-}$ are all central atoms and are cited first. The subsidiary atoms then follow, in alphabetical order of symbols (this rule is slightly modified for coordination compounds).

Examples

1. $[CrO_7S]^{2-}$	5. H_3PO_4
2. $[ICl_4]^-$	6. $SbCl_2F$
3. ClO^-	7. $PBrCl_2$
4. NO_2^-	

Slightly different rules apply to coordination compounds, the molecules (or, when charged, complex ions) of which are considered to be composed of a central atom to which are coordinated ligands by (to a first approximation) donor–acceptor electron-pair bonds. The ligands are grouped as formally anionic or formally neutral. The anionic ligands are cited first (alphabetical order of first symbols) and the neutral ligands next (also in alphabetical order of first symbols). The whole coordination entity (which may be positive, negative or neutral) is enclosed in square brackets.

Organic ligands are cited under C, and NO and CO are regarded as neutral. Because square brackets are always of highest seniority (or priority), a hierarchical sequence of enclosing marks is adopted to ensure that this seniority is preserved: [ ], [( )], [{( )}], [{[( )]}], [{{[( )]}}], etc.

Table 3.2 Some important compound classes and functional groups.

Class	Functional group	General constitution*	
Alkanes	None	C_nH_{2n+2}	
Alkenes	C=C	R_2C=CR_2	(R or Ar or H)
Alkynes	C≡C	RC≡CR	(R or Ar or H)
Alcohols	-OH	R-OH	
Aldehydes	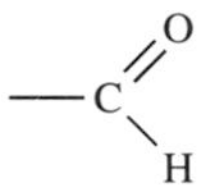	R-CHO	(R or Ar)
Amides	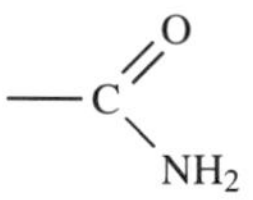	$R\text{-}CONH_2$	(R or Ar)
Amines	$\text{-}NH_2$, -NHR, $\text{-}NR_2$	$R\text{-}NH_2$ R-NH-R $R\text{-}NR_2$	(R or Ar)
Carboxylic acids	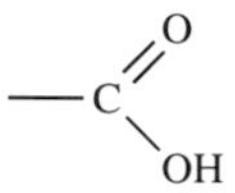	R-COOH	(R or Ar)
Ethers	-O-	R-O-R	(R or Ar)
Esters	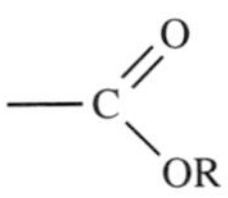	R-COOR	(R or Ar)
Halogeno compounds	-F, -Cl, -Br, -I	R-F, R-Cl R-Br, R-I	(R or Ar)
Ketones	>C=O	R-CO-R	(R or Ar)
Nitriles	-CN	R-CN	(R or Ar)

* In this table, and in common organic usage, Ar represents an aromatic group rather than the element of atomic number 18, and R represents an aliphatic group.

Examples

8. $[IrHCl_2(C_5H_5N)(NH_3)]$
9. $K_3[Fe(CN)_6]$
10. $[Ru(NH_3)_5(N_2)]Cl_2$
11. $K_2[Cr(CN)_2O_2(O_2)(NH_3)]$
12. $[Cu\{OC(NH_2)_2\}_2Cl_2]$
13. $[ICl_4]^-$

It is often a matter of choice whether a species is regarded as a coordination entity or not. Thus, sulfate may be regarded as a complex of S^{VI} with four O^{2-} ligands. It would then be represented as $[SO_4]^{2-}$, but it is not considered generally necessary to use square brackets here. The position with regard to $[ICl_4]^-$ is not so clear-cut: $[ICl_4]^-$, $(ICl_4)^-$ and ICl_4^- would all be acceptable, depending upon the precise circumstances of use.

For certain species it is not possible to define a central atom. Thus, for chain species, such as thiocyanate, the symbols are cited in the order in which they appear in the chain.

Examples

14. -SCN   17. -NCS
15. HOCN   18. HCNO
16. $(O_3POSO_3)^-$

Addition compounds are represented by the formulae of the individual constituent species, with suitable multipliers that define the appropriate molecular ratios of the constituent species, and separated by centre dots. In general, the first symbols determine the order of citation.

Examples

19. $3CdSO_4 \cdot 8H_2O$
20. $8H_2S \cdot 46H_2O$
21. $BF_3 \cdot 2H_2O$

These suggestions are advisory and should be used where there are no overriding reasons why they should not be. For example, PCl_3O is a correct presentation but, because the group P=O persists in whole families of compounds, the presentation $POCl_3$ may be more useful in certain contexts. There is no objection to this.

The concept of a group is especially important in organic chemistry. A functional group represents a set of atoms that is closely linked with chemical reactivity and defined classes of substances. For instance, the functional group hydroxyl, -OH, is characteristic of the classes alcohol, phenol and enol. Alcohols are often represented by the general formula R-OH, in which R- represents a hydrocarbon group typical of aliphatic and alicyclic substances.

A functional group is a set of atoms that occurs in a wide range of compounds and confers upon them a common kind of reactivity (see Table 3.2). Phenols are generally represented by Ar-OH, in which Ar- represents an aromatic skeleton, composed of benzene rings or substituted benzene rings. Enols are molecules in which the -OH group is linked to an atom that is also engaged in a double bond.

Examples

Typical alcohols

22.   23.

Typical phenols

24. 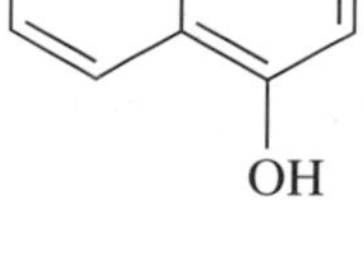   25. 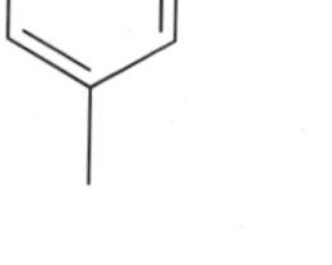   26.

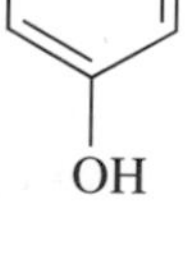

A typical enol

27.

The formulae discussed so far rely on a minimum of structural information. Increasingly, chemists need to convey more than a list of constituents when providing a formula. They need to say something about structure; to do this, simple line formulae (i.e. formulae written on a single line, as is text) need to be modified. How they are modified is determined by what information needs to be conveyed. Sometimes this can take a simple modification of a line formula to show extra bonds not immediately apparent, as in ring compounds, either organic or coordination compounds.

Examples

28. [NiS═{P(CH$_3$)$_2$}(C$_5$H$_5$)]   29. ClCHCH$_2$CH$_2$CH$_2$CH$_2$CH$_2$

Note that these bond indicators do not imply long bonds. Their size and form are dictated solely by the demands of the linear presentation.

It is usual for a coordination compound to write the formula of a ligand with the donor atom first. The nickel complex represented above has both S and P bonded to the metal (as well as all the carbon atoms of the C$_5$H$_5$). The ring structure for chlorocyclohexane should be obvious.

However, in many cases it is not possible to show all the necessary detail in a line formula. In such cases, attempts must be made to represent structures in three dimensions.

3.7   THREE-DIMENSIONAL STRUCTURES AND PROJECTIONS

The approach adopted is to view the molecule in three dimensions, imagining each atom or group to be placed at a vertex of an appropriate polyhedron. In organic chemistry this is usually the tetrahedron with carbon at the centre. Table 3.3 (p. 18) shows the polyhedra normally encountered in organic and inorganic chemistry. It also includes for each polyhedron the polyhedral symbols to denote shape and coordination number. It is to be noted that these polyhedra are often presented in a highly formalised fashion. An octahedron is often represented with the apices rather than the octahedral faces depicted, thus:

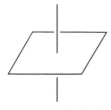

An octahedral complex, such as [Co(NH$_3$)$_3$(NO$_2$)$_3$], would have an acceptor at the central position and a ligand at each of the six apices, thus:

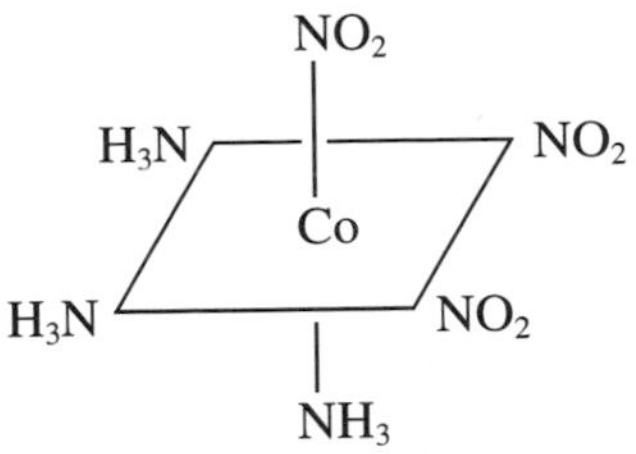

$$NO_2$$

$$H_3N - Co - NO_2$$ (structural diagram with NH₃ below)

This is *not* intended to indicate bonds between, for example, H_3N and NO_2, and it is perhaps an unfortunate hybrid of a three-dimensional representation and a line formula in which only selected bonds are shown. Care needs to be exercised when using this format, and it is not to be recommended, especially for beginning students. A more accurate and simpler representation is shown below.

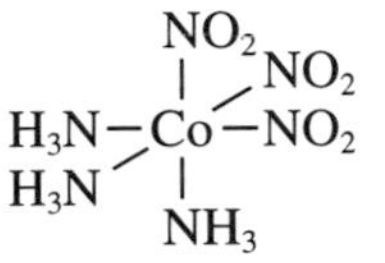

$$\begin{array}{c} NO_2 \\ | \quad NO_2 \\ H_3N - Co - NO_2 \\ H_3N \quad | \\ NH_3 \end{array}$$

Perspective can be enhanced by shaping the bonds directed out of the plane of the paper.

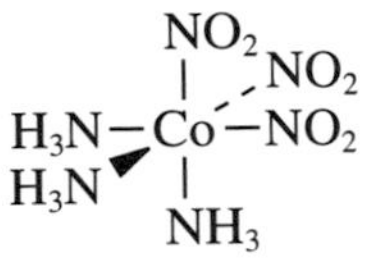

$$\begin{array}{c} NO_2 \\ | \quad NO_2 \\ H_3N - Co - NO_2 \\ H_3N \quad | \\ NH_3 \end{array}$$

Normally, a two-electron bond is represented in these formulae by a line. When electron pairs are not conveniently localised between specific atom pairs, it is not possible to represent bonds so. For example, benzene can be represented as

or, perhaps, more accurately

In complex compounds, similar representations are used:

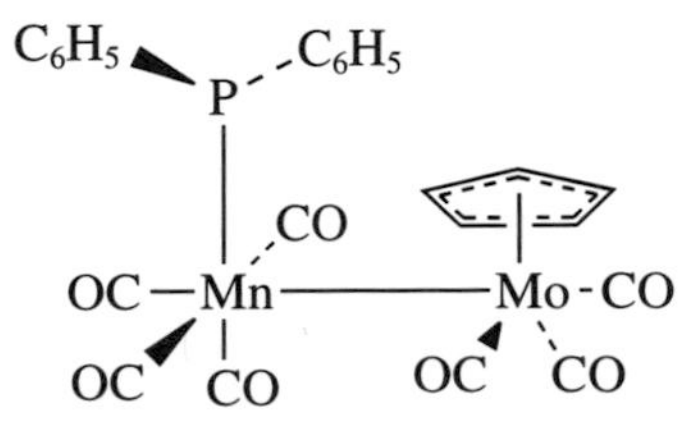

$$C_6H_5 \quad P \quad C_6H_5$$
$$OC - Mn \longrightarrow Mo - CO$$
$$OC \quad CO \qquad OC \quad CO$$

Projections are used, particularly in organic chemistry, to represent three-dimensional molecules in two dimensions. In a Fischer projection, the atoms or groups of atoms attached to a tetrahedral centre are projected onto the plane of the paper from such an orientation that atoms or groups appearing above or below the

Table 3.3 Polyhedral symbols and geometrical structures.

Polyhedra of four-coordination

tetrahedron

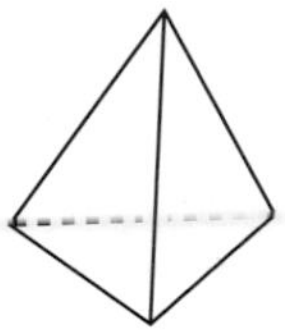

square planar

T-4

SP-4

Polyhedra of five-coordination

trigonal bipyramid

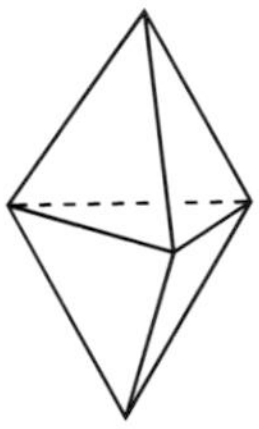

square pyramid

TBPY-5

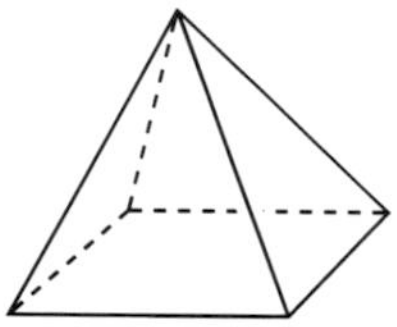

SPY-5

Polyhedra of six-coordination

octahedron

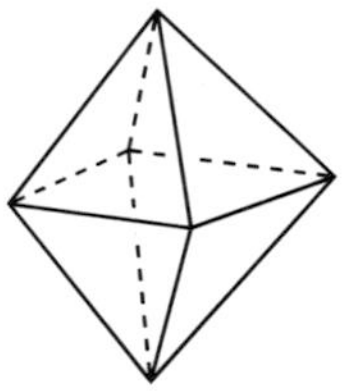

trigonal prism

OC-6

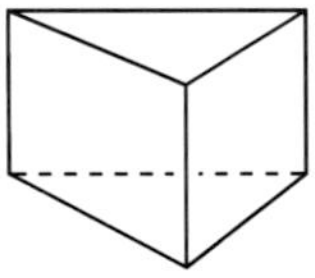

TPR-6

Polyhedra of seven-coordination

| pentagonal bipyramid | octahedron, face monocapped | trigonal prism, square face monocapped |

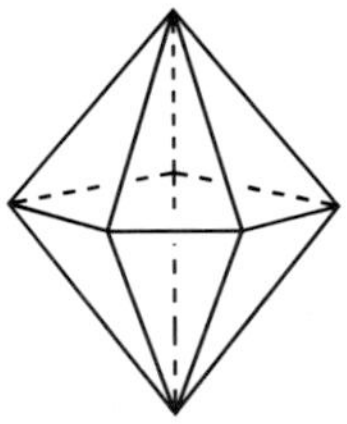

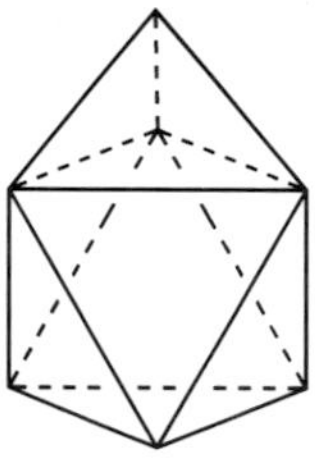

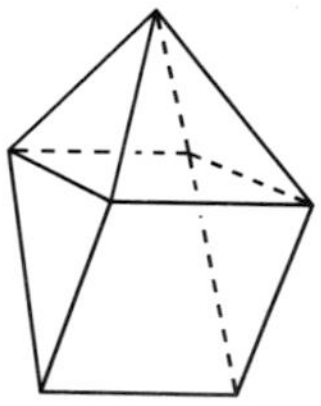

PBPY-7

OCF-7

TPRS-7

Polyhedra of eight-coordination

| cube | square antiprism | dodecahedron | hexagonal bipyramid |

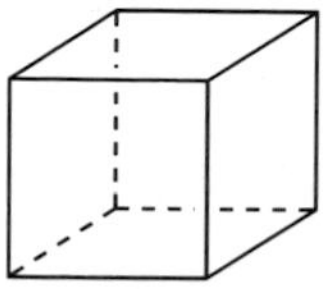

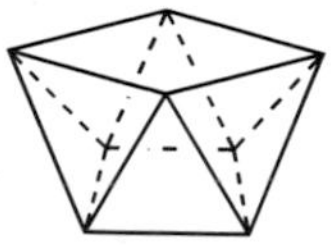

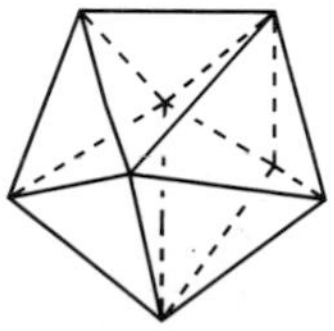

CU-8

SAPR-8

DD-8

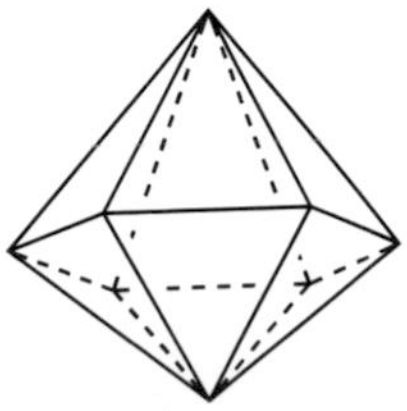

HBPY-8

Continued.

above for the octahedron, and the same caution should be used also with these representations. The central atom is represented here by the letter M and the attached groups by letters a, b, c, etc. For a given formula (e.g. Mabcde) more than one shape may be possible:

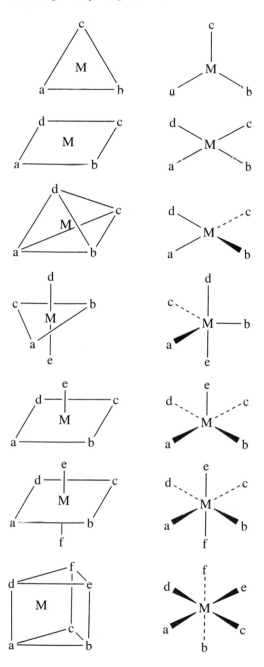

3.8   ISOMERS AND STEREOISOMERS

Isomerism describes the relationship between molecular entities having the same molecular formula, but differing in structure and/or connectivity between the constituent atoms. For example, the molecular formula C_7H_{16} corresponds to many different alkanes differing from each other in their connectivities. Two are shown in (a) below. In the same manner, two structural formulae can be envisaged for the molecular formula C_3H_6O, one belonging to the class of ketones, and the other being an alcohol (b).

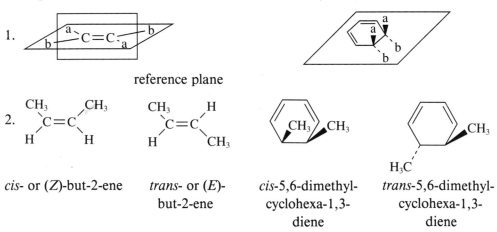

(a)   and

(b) CH_3-CO-CH_3 and CH_2=CH-CH_2-OH

Stereoisomers are isomers having the same molecular formula and the same connectivity, but different spatial arrangements. There are three classes of stereoisomer: *cis–trans* isomers, conformational isomers and enantiomers.

3.8.1   *cis–trans* Isomers

These are associated with tetrahedral and octahedral spatial distributions of atoms, and with bonds. The stereodescriptors *cis* and *trans* indicate the spatial distribution with reference to a plane defined by the molecular structure, often in relation to a double bond.

Examples

1.

reference plane

2.

cis- or (Z)-but-2-ene   trans- or (E)-but-2-ene   cis-5,6-dimethyl-cyclohexa-1,3-diene   trans-5,6-dimethyl-cyclohexa-1,3-diene

In many cases the *cis–trans* stereodescriptors are ambiguous and they are now often replaced by stereodescriptors E and Z, which represent the relative seniorities of the groups attached to the double bond. They are assigned using the Cahn–Ingold–Prelog (CIP) rules (see the *Guide to IUPAC Nomenclature of Organic Compounds*, pp. 151–154). This system of seniorities is based upon relative atomic numbers and is used in both organic and inorganic nomenclatures. For other organic systems of seniority, see Tables 4.10 and 6.1 and Chapter 4, Section 4.5.6 (p. 84).

The *cis–trans* stereodescriptors are acceptable for simple organic structures and they have been used also to describe spatial distribution in octahedral and square-planar structures. However, they are not adequate to distinguish all possibilities. The system that is currently recommended for complexes is described in more detail in the *Nomenclature of Inorganic Chemistry*, Chapter 10.

3.8.2 **Conformational isomers (or conformers)**

The conformation of a molecule is the spatial arrangement of the atoms. Different stereoisomers that can be interconverted by rotation about single bonds are termed conformers. Thus a conformer is one of a set of stereoisomers differing from one another in their conformations, each of which is considered to correspond to a potential-energy minimum. The interconversion of conformers by rotation around a single bond involves crossing an energy barrier between different potential-energy minima.

Examples

1.

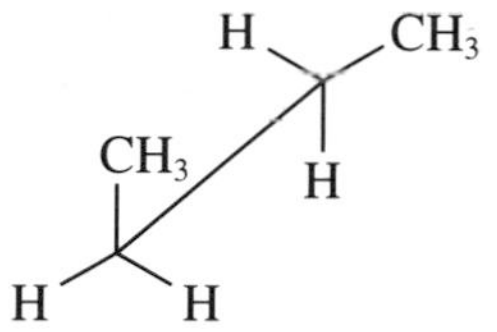

synclinal or gauche conformation

2.

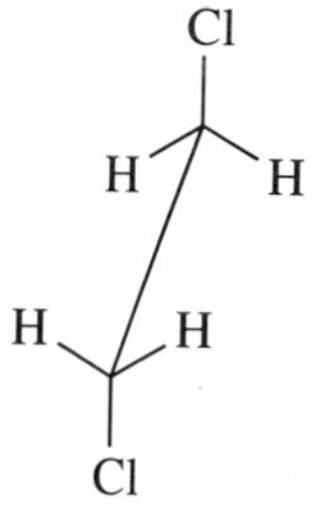

antiperiplanar conformation

The concept of conformational analysis has led to a better understanding of the spatial arrangements of cyclic alkanes and of the chemical reactivity of functionalized derivatives. A specific terminology is used.

Examples

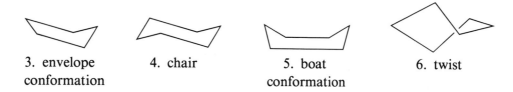

3. envelope conformation

4. chair

5. boat conformation

6. twist

In a cyclohexane or similar molecule, extraskeletal bonds are equatorial (b) or axial (a).

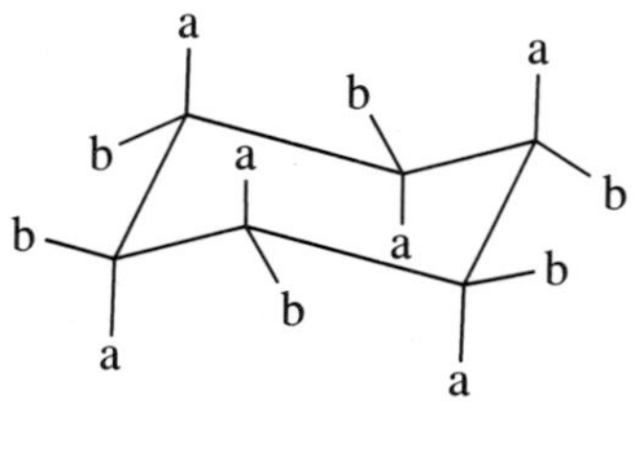

3.8.3 **Enantiomers**

Chirality is the property of an object that is not identical with its mirror image. For example, the human left hand has the same shape and internal structure as the human right hand, but they are different non-superimposable objects. They are mirror images of each other. Where two such molecules exist in chemistry, they are called enantiomers. Enantiomers have identical physical properties (except for the interaction with polarised light) and chemical reactivity (except for reaction with other chiral species). Consequently, any biological activities that involve stereospecificity may also be very different. The directions of the specific rotations are equal and opposite. A chiral molecule is not superimposable on its mirror image, whereas an achiral molecule is. Chirality is due to the presence in a molecule of a chiral centre, axis or plane. Only chiral centres will be dealt with here.

A chiral centre is an atom binding a set of ligands in a spatial arrangement that is not superimposable on its mirror image, e.g. a carbon compound Cabcd, a phosphorus compound Pabc and an ammonium ion $(Nabcd)^+$. The stereodescriptors R and S are used to describe each enantiomer. These are selected using the CIP priorities assigned to the substituents a, b, c and d by the methods described in the *Guide to IUPAC Nomenclature of Organic Compounds*, p. 152.

Example

Mirror plane

1.

$$H_2N \diagdown C \diagup COOH \quad \vert \quad HOOC \diagdown C \diagup NH_2$$
$$H \diagup \quad \diagdown CH_3 \quad \vert \quad H_3C \diagup \quad \diagdown H$$

Two enantiomers

R $\qquad\qquad$ S

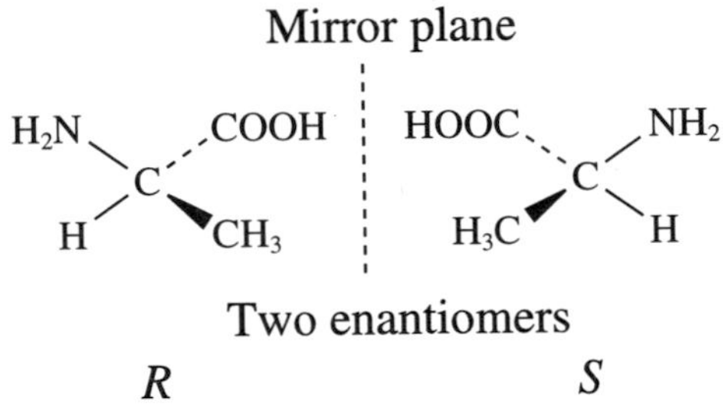

When a molecule contains two chiral centres, four or three stereoisomers are possible, depending on whether or not a plane of symmetry is present. In a set of four isomers, there are two pairs of enantiomers. Isomers that are not enantiomers are called diastereoisomers. In the following example, compounds I and II are enantiomers and compounds III and IV are enantiomers, but compounds I and III, for instance, are diastereoisomers.

Example

2.

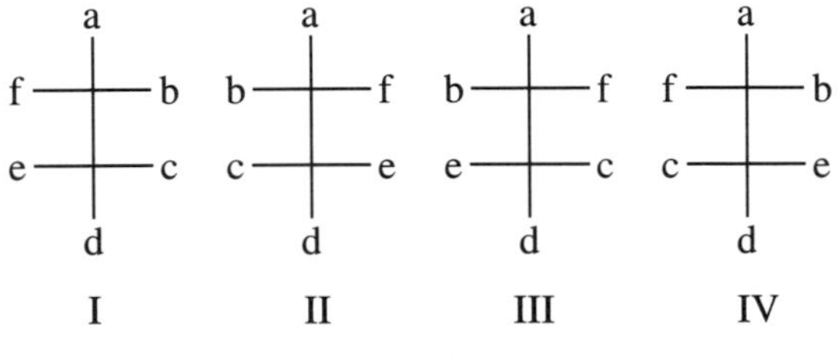

	a		a		a		a	
f	—	b	b	—	f	b	—	f	f	—	b
e	—	c	c	—	e	e	—	c	c	—	e
	d		d		d		d	
	I		II		III		IV	

A plane of symmetry makes the molecule achiral and thus reduces the number of isomers. The molecule below with the plane of symmetry is designated *meso*.

Example

Mirror plane

3.

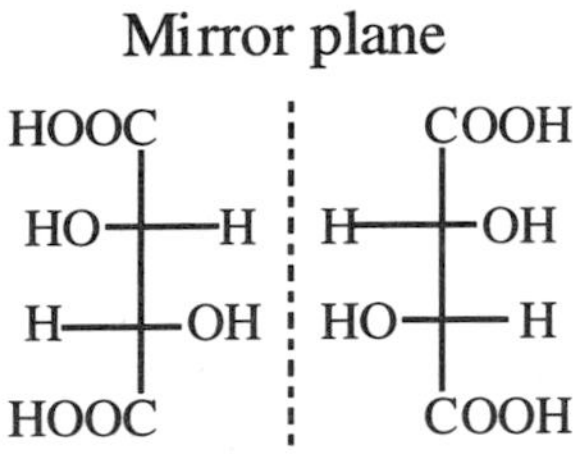

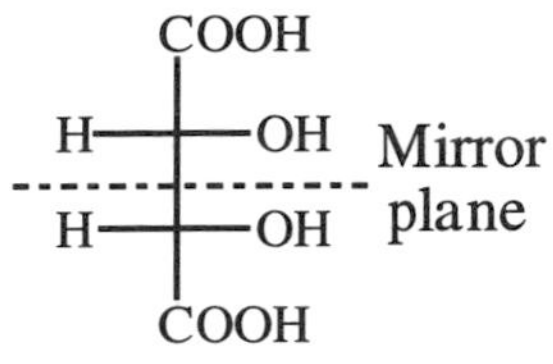

enantiomers   *meso* compound

A *meso* compound has a specific rotation of polarised light of 0°. A racemate is an equimolar mixture of two enantiomers and its specific rotation is also 0°.

The examples below further demonstrate the use in organic nomenclature of the stereodescriptors described above.

Examples

4.

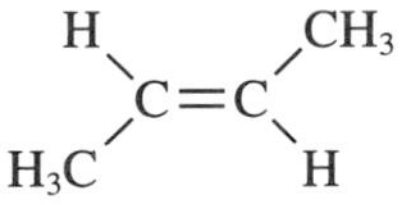

(*E*)-but-2-ene

5.

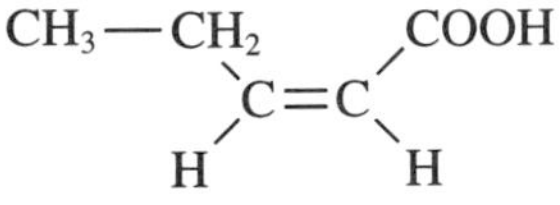

(*Z*)-pent-2-enoic acid

6.

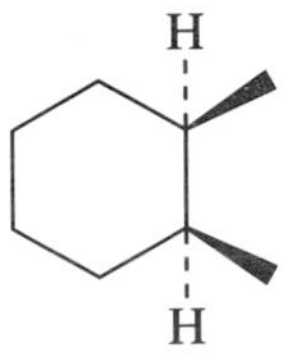

cis-1,2-dimethylcyclohexane

7.

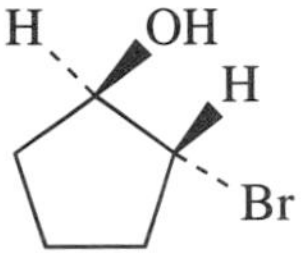

trans-2-bromocyclopentan-1-ol

8.

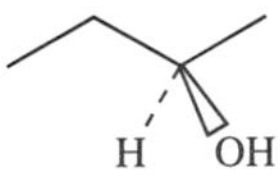

(*R*)-butan-2-ol

9.

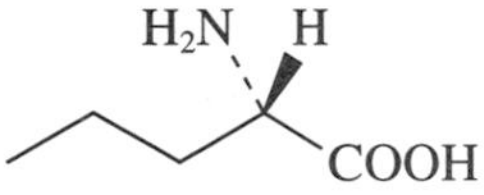

(*S*)-2-aminopentanoic acid

4 Naming of substances

TYPES OF NOMENCLATURE

Specialists in nomenclature recognise two different categories of nomenclature. Names that are arbitrary (including the names of the elements, such as sodium and hydrogen) as well as laboratory shorthand names (such as diphos and LithAl) are termed trivial names. This is not a pejorative or dismissive term. Trivial nomenclature contrasts with systematic nomenclature, which is an assembly of rules, themselves arbitrary. The function of specialists in nomenclature is to codify such rules so that everyone can use them to identify pure substances, rather like many of us use an alphabet to represent words. There may be more than one way to name a compound or species, and no one way may be superior to all the others. Names also vary in complexity, depending upon how much information needs to be conveyed. For example, a compositional name conveys less information than a structural (or constitutional) name, because this includes information about the arrangement of atoms in space.

Chemists have developed names for materials since the beginning of the science. Initially, the names were always trivial, because the systematics of molecular structure were completely unknown. The names of the elements are still essentially trivial, but these are the basis of systematic nomenclature.

Now that we understand much more of the way in which atoms combine, we can construct names that can give information about stoichiometry and structure. However, unsystematic usages that have their roots in the distant past are still embedded in the nomenclature. In addition, there are several systems of nomenclature, and these tend to reflect the kinds of chemistry for which they have been developed.

4.1.1 Binary-type nomenclature

This is a system based upon stoichiometry. It is not restricted to binary (two-element) compounds, but the nomenclature is binary in structure, as discussed below.

4.1.2 Coordination-type nomenclature

This system is additive and was developed originally to name coordination compounds, although it can be used in other circumstances when appropriate. For a discussion, see the *Nomenclature of Inorganic Chemistry*, Chapter 10. The compound to be named is considered as a central atom together with its ligands, and the name is developed by assembling the individual names of the constituents. This system has also been applied to name oxoacids and the related anions. Coordination names for oxoanions are cited in the examples throughout the text, and they are presented in detail in Section 4.4.5 (p. 69).

4.1.3   Substitutive-type nomenclature

This is the principal nomenclature system used in organic chemistry, as described in the *Guide to IUPAC Nomenclature of Organic Compounds*, p. 18. It is based upon the name of a formal parent molecule (normally a hydride), which is then substituted. Although it is principally an organic nomenclature, it has been extended to names of hydrides of Groups 14, 15, 16 and 17.

These systems may all be applied to the same compound. The name adopted is then a matter of choice or convenience. Thus, $SiCl_4$ can be named silicon tetrachloride (binary), tetrachlorosilicon (coordination) and tetrachlorosilane (substitutive). No one name is 'better' or 'more correct' than any other.

Other minor systems are also in use. Some are traditional, and some are very restricted in their application. These include acid nomenclature (inorganic, for oxoacids and derivatives), replacement nomenclature (mainly organic, to denote replacement of skeletal atoms in a parent rather than replacement of hydrogen atoms — oxa-aza replacement is one variant), functional class nomenclature (this is again principally organic and involves the use of type names such as alcohol, acid and ether) and subtractive nomenclatures (such as organic-deoxy and inorganic-debor). These will all be referred to briefly as appropriate.

4.2   BINARY-TYPE NOMENCLATURE

Although it is possible to develop a name based simply on an empirical formula (a reasonable proposal might be calcium sulfur tetraoxygen for $CaSO_4$), this is never done. Binary nomenclature is principally inorganic, and has no real simple organic counterpart.

4.2.1   Basis of the binary system

This provides names for compounds for which little or no structural information is available. However, a minimum of structural information is known or assumed. In particular, using the assumed or established division of constituents into positive and negative parts already employed above in establishing formulae, we divide the constituents into the same two classes, hence the term 'binary nomenclature'.

The positive and negative parts are sometimes referred to as electropositive and electronegative. However, there is no general scale of electropositiveness, and constituents are really more or less electronegative and are divided into groups of greater and lesser electronegativity. As discussed in Chapter 3 on formulae, even this is not to be interpreted with too much rigidity, and in nomenclature various arbitrary devices are used to define electronegativity. We shall continue to use the terms electropositive and electronegative because they are sanctioned by long nomenclature usage. In no circumstances should numerical values be applied to such terms.

4.2.2   Name derivation

The name is derived by combining the names of the electropositive constituent(s) with those of the electronegative constituent(s), suitably modified by any necessary multiplicative prefixes. The electropositive constituent names are cited first, and are separated from the electronegative constituent names by a space. The multiplicative prefixes may not be necessary if the oxidation states are explicit or are clearly understood. However, oxidation state information should never be conveyed by the suffixes -ous and -ic. This is confusing in the names of complexes (compare ferrous with cuprous and ferric with cupric, where the same suffix implies different oxidation states). The oxidation state should always be explicit and designated by roman numerals. Names of acids, such as sulfurous and nitrous, and sulfuric and nitric, present the same problem. Here, coordination names are also preferred and there are numerous examples throughout the text.

Examples
1. $NaCl$ sodium chloride
2. Ca_3P_2 calcium phosphide
3. Fe_3O_4 triiron tetraoxide

The name of the electropositive constituent is simply the unmodified element name, the name of a polyatomic cation or an accepted group name, as appropriate.

Examples
4. NH_4Cl ammonium chloride
5. UO_2Cl_2 uranyl dichloride
6. $O_2[PtF_6]$ dioxygen hexafluoroplatinate
7. OF_2 oxygen difluoride
8. $NOHSO_4$ nitrosyl hydrogensulfate

If there is more than one electropositive constituent, the names should be spaced and cited in alphabetical order of the initial letters, or of the second letters if the first letters are the same. Multiplicative prefixes are ignored for purposes of ordering.

Hydrogen is an exception. It is always cited last among the electropositive constituents and is separated from the following anion names by a space unless it is known to be bound to the anion. In languages other than English, different ordering may apply. In the examples, the letters defining the order are in bold face for clarity. This should not be extended to normal practice.

Examples
9. $KMgCl_3$ **m**agnesium **p**otassium chloride
10. $NaNH_4HPO_4$ **a**mmonium **s**odium hydrogenphosphate
11. $Cs_3Fe(C_2O_4)_3$ tric**a**esium **i**ron tris(oxalate)
12. $AlK(SO_4)_2 \cdot 12H_2O$ **a**luminium **p**otassium bis(sulfate)—water(1/12)

This last example shows how the formula of a compound considered as an addition compound is converted to a name. The molecular proportions are shown as the appropriate ratio (here, 1/12) in parentheses after the names, which are separated by a long dash.

The names of monoatomic electronegative constituents are derived from the names of the elements, but modified. The termination is replaced by the anion designator -ide. The reason why the treatment of these names is different from that for electropositive constituents is historical, and has no obvious logical basis. In certain cases, the modification is accompanied by an abbreviation and there are a few anion names that are based on Latin roots, although the element names are based on English. All these names are given in Table 4.1.

If there is more than one electronegative constituent the names are ordered alphabetically, as with the electropositive names.

Examples

13.	KCl	potassium chloride
14.	$BBrF_2$	boron bromide difluoride
15.	PCl_3O	phosphorus trichloride oxide
16.	$Na_2F(HCO_3)$	disodium fluoride hydrogencarbonate

Note that in the last example, 'disodium' is equally as acceptable as 'sodium', but in most circumstances the di- would be assumed to be obvious. The name hydrogencarbonate (no space) implies that the hydrogen is bonded in some fashion to the carbonate fragment. The presence of a space would imply that it is not.

The names of polyatomic electronegative groups are derived in various ways. Homoatomic species are designated using an appropriate multiplicative prefix.

Examples

17.	$(Sn_9)^{4-}$	nonastannide
18.	$(I_3)^-$	triiodide
19.	S_2^{2-}	disulfide

Note that upon forming a full name — as in Na_4Sn_9, tetrasodium (nonastannide) and $Tl(I_3)$, thallium (triiodide) — enclosing marks may be useful to avoid ambiguity. Compare with TlI_3, thallium triiodide. In some circumstances, such as when the name of the electronegative species is cited alone, e.g. in the discussion of an anion, it may be useful to indicate the charge: Sn_9^{4-}, nonastannide(4–); $(I_3)^-$, triiodide(1–); S_2^{2-}, disulfide(2–). This is discussed further below. Some trivial names are still allowed.

Examples

20.	O_2^-	superoxide
21.	O_2^{2-}	peroxide
22.	O_3^-	ozonide
23.	N_3^-	azide
24.	C_2^{2-}	acetylide

The names of heteroatomic electronegative constituents generally take the anion ending -ate, which is also characteristic of the names of anions of oxoacids (sulfate, phosphate, nitrate, etc.). Many such anions are coordination compounds, and these names are assembled using the rules of coordination nomenclature (see Section 4.4, p. 51).

Table 4.1 Names of ions and groups.*

Neutral atom or group formula	Name — Uncharged (atom, molecule or radical)	Cation or cationic group	Anion	Ligand
Ag	silver	silver	argentide	
Al	aluminium	aluminium	aluminide	
As	(mono)arsenic	arsenic	arsenide	arsenido
AsH_4		AsH_4^+ arsonium		
AsO_3			AsO_3^{3-} arsenite trioxoarsenate(3−) trioxoarsenate(III)	arsenito(3−) trioxoarsenato(3−) trioxoarsenato(III)
AsO_4			AsO_4^{3-} arsenate tetraoxoarsenate(3−) tetraoxoarsenate(V)	arsenato(3−) tetraoxoarsenato(3−) tetraoxoarsenato(V)
AsS_4			AsS_4^{3-} tetrathioarsenate(3−) tetrathioarsenate(V)	tetrathioarsenato(3−) tetrathioarsenato(V)
Au	gold	Au^+ gold(1+) gold(I) Au^{3+} gold(3+) gold(III)	auride	
B	(mono)boron	boron	boride	borido
BO_2			$(BO_2^-)_n$ metaborate poly[dioxoborate(1−)] poly[dioxoborate(III)]	metaborato
BO_3			BO_3^{3-} borate trioxoborate(3−) trioxoborate(III)	borato trioxoborato(3−) trioxoborato(III)

Continued on p. 32.

Formula	Name	Substituent group	Anion / acid	Ligand
Ba	barium	barium	baride	
Be	beryllium	beryllium	beryllide	
Br	(mono)bromine	bromine	bromide	bromo
BrO	bromine monoxide	bromosyl	BrO⁻ oxobromate(1–) oxobromate(i) (not hypobromite)	oxobromato(1–) oxobromato(i)
BrO₂	bromine dioxide	bromyl	BrO₂⁻ dioxobromate(1–) dioxobromate(iii) (not bromite)	dioxobromato(1–) dioxobromato(iii)
BrO₃	bromine trioxide	perbromyl	BrO₃⁻ trioxobromate(1–) trioxobromate(v) (not bromate)	trioxobromato(1–) trioxobromato(v)
BrO₄	bromine tetraoxide		BrO₄⁻ tetraoxobromate(1–) tetraoxobromate(vii) (not perbromate)	tetraoxobromato(1–) tetraoxobromato(vii)
Br₃	tribromine		tribromide(1–)	tribromo(1–)
C	(mono)carbon	carbon	carbide	carbido
CN			CN⁻ cyanide	cyano
CO	carbon monoxide	carbonyl		carbonyl carbon monoxide
CO₃			CO₃²⁻ carbonate trioxocarbonate(2–) trioxocarbonate(iv)	carbonato trioxocarbonato(2–) trioxocarbonato(iv)
CS	carbon monosulfide	thiocarbonyl		thiocarbonyl carbon monosulfide
CS₃			CS₃²⁻ trithiocarbonate(2–) trithiocarbonate(iv)	trithiocarbonato(2–) trithiocarbonato(iv)
C₂	dicarbon		C₂²⁻ acetylide dicarbide(2–)	dicarbido
Cl	(mono)chlorine	chlorine	chloride	chloro

31

Table 4.1 (*Continued.*)

Neutral atom or group formula	Name — Uncharged (atom, molecule or radical)	Cation or cationic group	Anion	Ligand
ClF_4	chlorine tetrafluoride	ClF_4^+ tetrafluorochlorine(1+) tetrafluorochlorine(v)	ClF_4^- tetrafluorochlorate(1−) tetrafluorochlorate(iii)	tetrafluorochlorato(1−) tetrafluorochlorato(iii)
ClO	chlorine monoxide	chlorosyl	ClO^- hypochlorite oxochlorate(1−) oxochlorate(i)	hypochlorito oxochlorato(1−) oxochlorato(i)
ClO_2	chlorine dioxide	chloryl	ClO_2^- chlorite dioxochlorate(1−) dioxochlorate(iii)	chlorito dioxochlorato(1−) dioxochlorato(iii)
ClO_3	chlorine trioxide	perchloryl	ClO_3^- chlorate trioxochlorate(1−) trioxochlorate(v)	chlorato trioxochlorato(1−) trioxochlorato(v)
ClO_4	chlorine tetraoxide		ClO_4^- perchlorate tetraoxochlorate(1−) tetraoxochlorate(vii)	perchlorato tetraoxochlorato(1−) tetraoxochlorato(vii)
Cm	curium	curium	curide	
Co	cobalt	Co^{2+} cobalt(2+) cobalt(ii) Co^{3+} cobalt(3+) cobalt(iii)	cobaltide	
Cr	chromium	Cr^{2+} chromium(2+) chromium(ii) Cr^{3+} chromium(3+) chromium(iii)	chromide	
CrO_2	chromium dioxide	chromyl		

Formula	Name (uncharged)	Cation	Anion	Ligand
CrO_4			CrO_4^{2-} chromate tetraoxochromate(2−) tetraoxochromate(vi)	chromato tetraoxochromato(2−) tetraoxochromato(vi)
Cr_2O_7			$Cr_2O_7^{2-}$ dichromate(2−) μ-oxo-hexaoxodichromate(2−) μ-oxo-hexaoxodichromate(vi)	dichromato(2−) μ-oxo-hexaoxodichromato(2−) μ-oxo-hexaoxodichromato(vi)
Cu	copper	Cu^+ copper(1+) copper(i) Cu^{2+} copper(2+) copper(ii)	cupride	
F	(mono)fluorine		fluoride	fluoro
Fe	iron	Fe^{2+} iron(2+) iron(ii) Fe^{3+} iron(3+) iron(iii)	ferride	
H	(mono)hydrogen		hydride	hydrido hydro (in boron compounds)
HCO_3			HCO_3^- hydrogencarbonate(1−) hydrogentrioxocarbonate(1−) hydrogentrioxocarbonate(iv)	hydrogencarbonato(1−) hydrogentrioxocarbonato(1−) hydrogentrioxocarbonato(iv)
HO	HO hydroxyl	HO^+ hydroxylium	OH^- hydroxide	hydroxido hydroxo
HO_2	hydrogen dioxide	hydrogenperoxide(1+)	HO_2^- hydrogenperoxide(1−) hydrogendioxide(1−)	hydrogenperoxo
HPO_4			HPO_4^{2-} hydrogenphosphate(2−) hydrogentetraoxophosphate(2−) hydrogentetraoxophosphate(v)	hydrogenphosphato(2−) hydrogentetraoxophosphato(2−) hydrogentetraoxophosphato(v)

Continued on p. 34.

Table 4.1 (*Continued.*)

Neutral atom or group formula	Name — Uncharged (atom, molecule or radical)	Cation or cationic group	Anion	Ligand
HS			HS⁻ hydrogensulfide(1−)	hydrogensulfido(1−) sulfanido
HSO₃			HSO₃⁻ hydrogensulfite(1−) hydrogentrioxosulfate(1−) hydrogentrioxosulfate(IV)	hydrogensulfito(1−) hydrogentrioxosulfato(1−) hydrogentrioxosulfato(IV)
HSO₄			HSO₄⁻ hydrogensulfate(1−) hydrogentetraoxosulfate(1−) hydrogentetraoxosulfate(VI)	hydrogensulfato(1−) hydrogentetraoxosulfato(1−) hydrogentetraoxosulfato(VI)
H₂O	oxidane water			aqua oxidane
H₃O	trihydrogen oxide	H₃O⁺ oxonium		
H₂PO₄			H₂PO₄⁻ dihydrogenphosphate(1−) dihydrogentetraoxophosphate(1−) dihydrogentetraoxophosphate(V)	dihydrogenphosphato(1−) dihydrogentetraoxophosphato(1−) dihydrogentetraoxophosphato(V)
Hg	mercury	Hg²⁺ mercury(2−) mercury(II) Hg₂²⁺ dimercury(2+) dimercury(I)	mercuride	
I	(mono)iodine	iodine	iodide	iodo
IF₄	iodine tetrafluoride	IF₄⁺ tetrafluoroiodine(1+) tetrafluoroiodine(V)	IF₄⁻ tetrafluoroiodate(1−) tetrafluoroiodate(III)	tetrafluoroiodato(1−) tetrafluoroiodato(III)
IO	iodine oxide	iodosyl	IO⁻ oxoiodate(1−) oxoiodate(I) (not hypoiodite)	oxoiodato(1−) oxoiodato(I)

IO_2	iodine dioxide	iodyl	IO_2^- dioxoiodate(1−) dioxoiodate(III) (not iodite)	dioxoiodato(1−) dioxoiodato(III)
IO_3	iodine trioxide	periodyl	IO_3^- iodate trioxoiodate(1−) trioxoiodate(V)	iodato trioxoiodato(1−) trioxoiodato(V)
IO_4	iodine tetraoxide		IO_4^- periodate tetraoxoiodate(1−) tetraoxoiodate(VII)	periodato tetraoxoiodato(1−) tetraoxoiodato(VII)
IO_6			IO_6^{5-} hexaoxoiodate(5−) hexaoxoiodate(VII)	hexaoxoiodato(5−) hexaoxoiodato(VII)
I_3	triiodine		triiodide(1−)	triiodo(1−)
In	indium	indium	indide	
Ir	iridium	iridium	iridide	
K	potassium	potassium	kalide	
Li	lithium	lithium	lithide	
Mg	magnesium	magnesium	magneside	
Mn	manganese	Mn^{2+} manganese(2+) manganese(II) Mn^{3+} manganese(3+) manganese(III)	manganide	
MnO_4			MnO_4^- permanganate tetraoxomanganate(1−) tetraoxomanganate(VII) MnO_4^{2-} manganate tetraoxomanganate(2−) tetraoxomanganate(VI)	permanganato tetraoxomanganato(1−) tetraoxomanganato(VII) manganato tetraoxomanganato(2−) tetraoxomanganato(VI)
Mo	molybdenum	molybdenum	molybdenum	
N	(mono)nitrogen	nitrogen	nitride	nitrido

Continued on p. 36.

Table 4.1 (*Continued.*)

Neutral atom or group formula	Name — Uncharged (atom, molecule or radical)	Cation or cationic group	Anion	Ligand
NCO (see OCN)				
NH			NH^{2-} imide / azanediide / azanide(2−)	imido / azanediido
NHOH			$NHOH^-$ hydroxyamide	hydroxyamido
NH_2			NH_2^- amide / azanide	amido / azanido
NH_3	azane / ammonia	NH_3^+ ammoniumyl / azaniumyl		ammine / azane
NH_4		NH_4^+ ammonium / azanium		
NO	nitrogen monoxide	nitrosyl	NO^- oxonitrate(1−) / oxonitrate(I)	nitrosyl / nitrogen monoxide
NO_2	nitrogen dioxide	nitryl / nitroyl	NO_2^- nitrite	nitro / nitrito-O / nitrito-N
			dioxonitrate(1−) / dioxonitrate(III) / NO_2^{2-} dioxonitrate(2−) / dioxonitrate(II) (not nitroxylate)	dioxonitrato(1−) / dioxonitrato(III) / dioxonitrato(2−) / dioxonitrato(II)
NO_3	nitrogen trioxide		NO_3^- nitrate / trioxonitrate(1−) / trioxonitrate(V)	nitrato / trioxonitrato(1−) / trioxonitrato(V)

Formula				
N₂H		N₂H⁺ diazynium	N₂H⁻ diazenide N₂H³⁻ diazanetriide, diazanide(3–), hydrazinetriide, hydrazide(3–)	diazenido diazanetriido, hydrazido(3–)
N₂H₂	diazene, diimide	N₂H₂²⁺ diazynediium, diazynium(2+)	N₂H₂²⁻ diazanediide, hydrazide(2–), diazanide(2–), hydrazinediide	diazenediido, hydrazido(2–) N₂H₂ diazene, diimide
NHNH₂		N₂H₃⁺ diazenium	N₂H₃⁻ hydrazide, diazanide, hydrazinide	hydrazido, diazanido
N₂H₄	diazane, hydrazine	N₂H₄²⁺ diazenediium, diazenium(2+)		hydrazine, diazane
N₂H₅		N₂H₅⁺ hydrazinium(1+), diazanium		hydrazinium
N₂H₆		N₂H₆²⁺ hydrazinium(2+), diazanedium, diazanium(2+), hydrazinediium		
N₂O₂	dinitrogen dioxide		N₂O₂²⁻ dioxodinitrate(N–N)(2–), dioxodinitrate(N–N)(i) (not hyponitrite)	dioxodinitrato(N–N)(2–), dioxodinitrato(N–N)(i)
N₃	trinitrogen	trinitrogen	azide, trinitride(1–)	azido, trinitrido(1–)
Na	sodium	sodium	natride	

Continued on p. 38.

Table 4.1 (*Continued.*)

Neutral atom or group formula	Name Uncharged (atom, molecule or radical)	Cation or cationic group	Anion	Ligand
Ni	nickel	Ni^{2+} nickel(2+) nickel(II) Ni^{3+} nickel(3+) nickel(III)	nickelide	
O	(mono)oxygen	oxygen	oxide	oxo oxido
OCN			cyanate	cyanato cyanato-*O* cyanato-*N*
			nitridooxocarbonate(1−) nitridooxocarbonate(IV)	nitridooxocarbonato(1−) nitridooxocarbonato(IV)
OH (see HO)				
ONC			fulminate carbidooxonitrate(1−) carbidooxonitrate(V)	fulminato carbidooxonitrato(1−) carbidooxonitrato(V)
O_2	dioxygen	O_2^{+} dioxygen(1+)	O_2^{2-} peroxide dioxide(2−) O_2^{-} hyperoxide superoxide dioxide(1−)	peroxo dioxido(2−) hyperoxo superoxido dioxido(1−) O_2 dioxygen
O_3	trioxygen ozone		O_3^{-} ozonide trioxide(1−)	ozonido trioxido(1−) O_3 trioxygen
Os	osmium	osmium	osmide	

Formula	Name	Cation	Anion	Ligand
P	(mono)phosphorus	phosphorus	P³⁻ phosphide	phosphido
PCl₄	phosphorus tetrachloride	PCl₄⁺ tetrachlorophosphonium / tetrachlorophosphonium(v) / tetrachlorophosphorus(1+) / tetrachlorophosphorus(v) / tetrachlorophosphanium(1+)	PCl₄⁻ tetrachlorophosphate(1−) / tetrachlorophosphate(iii)	tetrachlorophosphato(1−) / tetrachlorophosphato(iii)
PHO₃			PHO₃²⁻ phosphonate / hydridotrioxophosphate(2−)	phosphonato(2−) / hydridotrioxophosphato(2−)
PH₂O₂			PH₂O₂⁻ phosphinate / dihydridodioxophosphate(1−)	phosphinato / dihydridodioxophosphato(1−)
PH₄		PH₄⁺ phosphonium		
PO	phosphorus monoxide	phosphoryl		
PO₃			PO₃³⁻ phosphite / trioxophosphate(3−) / trioxophosphate(iii) / (PO₃⁻)ₙ metaphosphate / poly[trioxophosphate(1−)] / poly[trioxophosphate(v)]	phosphito(3−) / trioxophosphato(3−) / trioxophosphato(iii)
PO₄			PO₄³⁻ phosphate / orthophosphate / tetraoxophosphate(3−) / tetraoxophosphate(v)	phosphato(3−) / orthophosphato / tetraoxophosphato(3−) / tetraoxophosphato(v)
P₂O₇	diphosphorus heptaoxide		P₂O₇⁴⁻ diphosphate(4−) / μ-oxo-hexaoxodiphosphate(4−) / μ-oxo-hexaoxodiphosphate(v)	diphosphato(4−) / μ-oxo-hexaoxodiphosphato(4−) / μ-oxo-hexaoxodiphosphato(v)
Pb	lead	Pb²⁺ lead(2+) / lead(ii) / Pb⁴⁺ lead(4+) / lead(iv)	plumbide	

Continued on p. 40.

Table 4.1 (*Continued.*)

Neutral atom or group formula	Name Uncharged (atom, molecule or radical)	Cation or cationic group	Anion	Ligand
Pd	palladium	Pd^{2+} palladium(2+) palladium(II) Pd^{4+} palladium(4+) palladium(IV)	palladide	
Pt	platinum	Pt^{2+} platinum(2+) platinum(II) Pt^{4+} platinum(4+) platinum(IV)	platinide	
Rb	rubidium	rubidium	rubidide	
Re	rhenium	rhenium	rhenide	
ReO_4			ReO_4^- tetraoxorhenate(1−) tetraoxorhenate(VII) (not perrhenate) ReO_4^{2-} tetraoxorhenate(2−) tetraoxorhenate(VI) (not rhenate)	tetraoxorhena-o(1−) tetraoxorhena-o(VII) tetraoxorhenato(2−) tetraoxorhenato(VI)
S	(mono)sulfur	sulfur	sulfide	sulfido thio
SCN			thiocyanate nitridothiocarbonate(1−) nitridothiocarbonate(IV)	thiocyanato-*N* thiocyanato-*S* nitridothiocarbonato(1−) nitridothiocarbonato(IV)
SO	sulfur monoxide	sulfinyl thionyl		sulfur monoxide
SO_2	sulfur dioxide	sulfonyl sulfuryl	SO_2^{2-} dioxosulfate(2−) dioxosulfate(II) (not sulfoxylate)	dioxosulfato(2−) dioxosulfato(II) SO_2 sulfur dioxide

Formula	Name	Cation/acyl name	Anion formula	Anion name	Ligand name
SO_3	sulfur trioxide		$SO_3{}^{2-}$	sulfite / trioxosulfate(2–) / trioxosulfate(IV)	sulfito / trioxosulfato(2–) / trioxosulfato(IV)
SO_4	sulfur tetraoxide		$SO_4{}^{2-}$	sulfate / tetraoxosulfate(2–) / tetraoxosulfate(VI)	sulfato / tetraoxosulfato(2–) / tetraoxosulfato(VI)
SO_5			$SO_5{}^{2-}$	trioxoperoxosulfate(2–) / trioxoperoxosulfate(VI) (not peroxomonosulfate)	
S_2	disulfur		$S_2{}^{2-}$	disulfide(2–)	disulfido(2–)
S_2O_3	disulfur trioxide		$S_2O_3{}^{2-}$	thiosulfate / trioxothiosulfate(2–) / trioxothiosulfate(VI)	thiosulfato / trioxothiosulfato(2–) / trioxothiosulfato(VI)
S_2O_4			$S_2O_4{}^{2-}$	dithionite / tetraoxodisulfate(S–S)(2–) / tetraoxodisulfate(S–S)(III)	dithionito / tetraoxodisulfato(S–S)(2–) / tetraoxodisulfato(S–S)(III)
S_2O_5	disulfur pentaoxide	disulfuryl	$S_2O_5{}^{2-}$	μ-oxo-tetraoxodisulfate(2–) / μ-oxo-tetraoxodisulfate(IV) (not disulfite)	
S_2O_7			$S_2O_7{}^{2-}$	disulfate(2–) / μ-oxo-hexaoxodisulfate(2–) / μ-oxo-hexaoxodisulfate(VI)	disulfato(2–) / μ-oxo-hexaoxodisulfato(2–) / μ-oxo-hexaoxodisulfato(VI)
S_2O_8			$S_2O_8{}^{2-}$	μ-peroxo-hexaoxodisulfate(2–) / μ-peroxo-hexaoxodisulfate(VI) (not peroxodisulfate)	
Sb	(mono)antimony / antimony			antimonide	antimonido
SbH_4		$SbH_4{}^{+}$ stibonium			

Continued on p. 42.

Table 4.1 (*Continued.*)

Neutral atom or group formula	Name — Uncharged (atom, molecule or radical)	Name — Cation or cationic group	Anion	Ligand
SeO_4			SeO_4^{2-} tetraoxoselenate(2−) tetraoxoselenate(vi) (not selenate)	tetraoxoselenato(2−) tetraoxoselenato(vi)
Si	(mono)silicon	silicon	silicide	silicido
SiO_3			$(SiO_3^{2-})_n$ metasilicate poly[trioxosilicate(2−)] poly[trioxosilicate(iv)]	
SiO_4			SiO_4^{4-} orthosilicate tetraoxosilicate(4−) tetraoxosilicate(iv)	
Si_2O_7			$Si_2O_7^{6-}$ μ-oxo-hexaoxo-disilicate(6−) μ-oxo-hexaoxo-disilicate(iv)	
Sn	tin	Sn^{2+} tin(2+) tin(ii) Sn^{4+} tin(4+) tin(iv)	stannide	
Te	(mono)tellurium	tellurium	telluride	tellurido
TeO_3			TeO_3^{2-} trioxotellurate(2−) trioxotellurate(iv)	
TeO_4			TeO_4^{2-} tetraoxotellurate(2−) tetraoxotellurate(vi)	
TeO_6			TeO_6^{6-} hexaoxotellurate(6−) hexaoxotellurate(vi) (not orthotellurate)	hexaoxotellurato(6−) hexaoxotellurato(vi)

Ti	titanium	titanium		titanide
TiO	titanium monoxide	oxotitanium(IV)		
Tl	thallium	thallium		thallide
U	uranium	uranium		uranide
UO_2	uranium dioxide	UO_2^+		
		uranyl(1+)		
		uranyl(V)		
		dioxouranium(1+)		
		dioxouranium(V)		
		UO_2^{2+}		
		uranyl(2+)		
		uranyl(VI)		
		dioxouranium(2+)		
		dioxouranium(VI)		
V	vanadium	vanadium		vanadide
VO	vanadium monoxide	oxovanadium(IV)		
W	tungsten	tungsten		tungstide
Zn	zinc	zinc		zincide
Zr	zirconium	zirconium		zirconide
ZrO	zirconium monoxide	oxozirconium(IV)		

* This table contains five columns, the first of which contains the symbol or formula of the neutral atom or group. The second column contains the corresponding name. The third column contains the name corresponding to the symbol or formula when it carries one or more units of positive charge. Inorganic nomenclature allows charges to be represented by the charge number, or to be inferred from an appropriate oxidation number. Both methods are displayed in the third column and in the succeeding columns. Formulae for ions are shown for cases where it is felt that confusion might otherwise arise. The fourth column contains the name of the symbol or formula when it carries one or more units of negative charge. Finally, the fifth column contains the name of the formula or symbol when the species it represents is a ligand (usually assumed to be anionic if it is not neutral).

The symbols (formulae) are listed in alphabetical order according to the principles outlined above. Because the terminations -ous and -ic for metal cation names are no longer recommended, these have been excluded, but we have attempted to include all those traditional names that are still allowed. We have not attempted to present names for species of very rare or unlikely occurrence, so there are gaps in the columns.

Users should note that we name only one specific structure for a given formula. In some cases there may be other structures that we have not named corresponding to that formula.

Examples
25. $[Fe(CO)_4]^{2-}$ tetracarbonylferrate(2–)
26. $[Cr(NCS)_4(NH_3)_2]^-$ diamminetetrathiocyanatochromate(1–)
27. SO_3^{2-} trioxosulfate(2–), preferred to sulfite
28. NO_2^- dioxonitrate(1–), preferred to nitrite

Some traditional names (often without -ate endings) are still allowed, although systematic coordination names are generally preferred.

Examples
29. CN^- cyanide
30. NH_2^- amide
31. OH^- hydroxide
32. AsO_3^{3-} arsenite
33. ClO_2^- chlorite
34. ClO^- hypochlorite
35. NO_2^- nitrite
36. SO_3^{2-} sulfite
37. $S_2O_4^{2-}$ dithionite

The order of citation within classes is always alphabetical. The use of multiplicative prefixes does not affect this order unless the prefix is part of the name: again, compare triiodide and (triiodide). Two sets of multiplicative prefixes are generally used (see Table 4.2). If two successive multiplicative prefixes are required, the Greek-based prefixes are recommended to be employed in the manner shown: $Ca(I_3)_2$, calcium bis(triiodide).

Where it is required to indicate oxidation state and/or charge, the former is indicated by using a roman numeral (in parentheses) as a suffix, and the latter by using an arabic numeral followed by the charge sign (all in parentheses) also as a suffix.

Table 4.2 Numerical prefixes.

1 Mono	19 Nonadeca
2 Di (bis)	20 Icosa
3 Tri (tris)	21 Henicosa
4 Tetra (tetrakis)	22 Docosa
5 Penta (pentakis)	23 Tricosa
6 Hexa (hexakis)	30 Triaconta
7 Hepta (heptakis)	31 Hentriaconta
8 Octa (octakis)	35 Pentatriaconta
9 Nona (nonakis)	40 Tetraconta
10 Deca (decakis), etc.	48 Octatetraconta
11 Undeca	50 Pentaconta
12 Dodeca	52 Dopentaconta
13 Trideca	60 Hexaconta
14 Tetradeca	70 Heptaconta
15 Pentadeca	80 Octaconta
16 Hexadeca	90 Nonaconta
17 Heptadeca	100 Hecta
18 Octadeca	

Examples

38. $UO_2{}^{2+}$ uranyl(VI) or dioxouranium(2+)
39. Na^- natride(−I)
40. $PO_4{}^{3-}$ phosphate(V)
41. N_2O nitrogen(I) oxide
42. Fe_3O_4 iron(II) diiron(III) tetraoxide
43. SF_6 sulfur(VI) fluoride
44. UO_2SO_4 uranyl(2+) sulfate or dioxouranium(VI) tetraoxosulfate(VI)
45. $(UO_2)_2SO_4$ uranyl(1+) sulfate or bis[dioxouranium(V)] tetraoxosulfate(VI)
46. Hg_2Cl_2 dimercury(I) chloride
47. $Fe_2(SO_4)_3$ iron(3+) sulfate or iron(III) sulfate or even diiron trisulfate

Qualification by both charge number and oxidation number is not allowed. It should be evident that there are several ways of conveying the same stoichiometric information, employing charge number, oxidation number and multiplicative prefixes. Employing them all would create redundancies. In general, one uses whichever devices are both necessary and sufficient, and no more.

The names described here can be used to develop further names with a little more manipulation. Addition compounds (a term that covers donor–acceptor complexes as well as a variety of lattice compounds) of uncertain structure can be named by citing the names of the constituent compounds and then indicating their proportions. Hydrates constitute a large class of compounds that can be represented by this means.

Examples

48. $3CdSO_4 \cdot 8H_2O$ cadmium sulfate—water(3/8)
49. $CaCl_2 \cdot 8NH_3$ calcium chloride—ammonia(1/8)
50. $BiCl_3 \cdot 3PCl_5$ bismuth(III) chloride—phosphorus(V) chloride(1/3)
51. $Al_2(SO_4)_3 \cdot K_2SO_4 \cdot 24H_2O$ aluminium sulfate—potassium sulfate—water(1/1/24)

Because the basic system described entails the consideration of the names of individual electropositive and electronegative substituents, it requires little elaboration (and this is implied in the text above) to use the same principles to name both anions and cations, and hence also salts. However, the binary names used here do not necessarily imply salt-like character.

The names of cations can be designated quite simply, although it is absolutely necessary to specify charge, either directly using charge number or indirectly using oxidation number. The parentheses in these examples of formulae are optional.

Examples

52. Na^+ sodium(1+) ion or sodium(I) cation
53. U^{6+} uranium(6+) ion or uranium(VI) cation
54. I^+ iodine(1+) ion or iodine(I) cation
55. $(O_2)^+$ dioxygen(1+) ion
56. $(Bi_5)^{4+}$ pentabismuth(4+) ion
57. $(Hg_2)^{2+}$ dimercury(2+) ion or dimercury(I) cation

Cations can also be obtained by the formal addition of a hydron (hydron is the recommended name for the normal isotopic mixture of protons, deuterons and tritons, see p. 7) to a binary hydride. In such cases, a formalism of substitutive nomenclature is used; the suffix -ium is added to the name, slightly modified, of the parent hydride. The selection of permitted hydride names and their usage are discussed in Section 4.5 on substitutive nomenclature.

Examples
58. H_3S^+ sulfanium
59. PH_4^+ phosphanium
60. SiH_5^+ silanium

Some variants are also allowed for mononuclear cations of Groups 15, 16 and 17. These are based on the usage of substitutive nomenclature, where the formal addition of a hydron to a parent hydride to give a cation is represented by the suffix -onium.

Examples
61. PH_4^+ phosphonium
62. AsH_4^+ arsonium
63. SbH_4^+ stibonium
64. H_3O^+ oxonium
65. H_3S^+ sulfonium
66. H_2I^+ iodonium

The name ammonium for NH_4^+ is not strictly systematic, but is hallowed by long usage and is therefore also allowed.

Derivatives of these hydrides, including organic derivatives, are named using the rules of substitutive nomenclature, or by using coordination nomenclature, as seems more appropriate.

Examples
67. $[PCl_4]^+$ tetrachlorophosphonium ion or tetrachlorophosphorus(1+)
68. $[P(CH_3)_2Cl_2]^+$ dichlorodimethylphosphonium ion or
                   dichlorodimethylphosphorus(1+)

Where the cation can clearly be regarded as a coordination complex, coordination nomenclature (see p. 51) is the natural choice.

Example
69. $[CoCl(NH_3)_5]^{2+}$ pentaamminechlorocobalt(2+) ion

These applications will be discussed further below.

There are a few special cases where trivial names are allowed. Some are listed here.

Examples

70.	NO^+	nitrosyl cation
71.	OH^+	hydroxylium cation
72.	NO_2^+	nitryl cation
73.	UO_2^{2+}	uranyl(2+) cation

The names of anions similarly are obtained by an extension of the names used for electronegative constituents, but with the proviso that the ending is always characteristic of an anion (ide, -ate or -ite, as discussed above). For the names of monoatomic anions, see Table 4.1. Homopolyatomic anions take names of the kind exemplified below.

Examples

74.	O_2^-	dioxide(1−)
75.	I_3^-	triiodide(1−)
76.	Pb_9^{4-}	nonaplumbide(4−)

Anions obtained formally by loss of a hydron from a parent hydride (see Table 5.2 p. 99 for a list of parent hydride names) are conveniently named by the methods of substitutive nomenclature.

Examples

77.	CH_3^-	methanide
78.	NH_2^-	amide or azanide
79.	PH^{2-}	phosphanediide or hydrogenphosphide(2−)
80.	SiH_3^-	silanide

Anions can also be formed by the loss of hydrons from acids. Where all the available hydrons are lost, the acid name is modified as shown below.

Examples

81. H_2SO_4, sulfuric acid → SO_4^{2-}, sulfate
82. H_3PO_4, phosphoric acid → PO_4^{3-}, phosphate
83. CH_3COOH, acetic acid → CH_3COO^-, acetate

In some cases, it may be necessary to add the charge to the name to distinguish different oxidation states or degrees of oxygenation (see also Section 4.4.5, p. 69). In most cases, coordination names are preferred for such species.

If not all the acid hydrons are lost, the situation is rather more complex.

Examples

84.	HSO_4^-	hydrogensulfate(1−)
85.	$H_2PO_4^-$	dihydrogenphosphate(1−)

The -ate termination is also used when hydrons are subtracted formally from an OH group in alcohols, etc.

Examples
86. CH_3O^- methanolate
87. $C_6H_5S^-$ benzenethiolate

An alternative way of envisaging hydride formation is by reaction of a neutral molecule with H^-, hydride ion. The -ate termination is again used.

Examples
88. BH_4^- tetrahydroborate
89. PH_6^- hexahydridophosphate
90. BCl_3H^- trichlorohydroborate

Note that certain oxoanions still retain trivial names and these are listed in Table 4.1. Note also the exceptional use of 'hydro' instead of the usual 'hydrido' to represent the bound hydride ion. This is restricted to boron nomenclature and survives for historical reasons.

Having established the methods for naming anions and cations, it is clear that salts have binary-type names that are often indistinguishable from the binary name assigned by dividing the constituents into electropositive and electronegative species. The names of cations always precede the names of anions in English and the names are always separated by spaces. This statement is true without exception. The orders of citation are alphabetical, with the exception of hydrogen among the cations, which is always cited last. Salts containing acid (replaceable) hydrogen contain the hydrogen name associated directly with its anion and without a space unless the hydron is unequivocally in cationic form.

Examples

91. $NaHCO_3$	sodium hydrogencarbonate
92. K_2HPO_4	dipotassium hydrogenphosphate
93. $KMgF_3$	magnesium potassium fluoride (optionally trifluoride)
94. $Na(UO_2)_3[Zn(H_2O)_6](CH_3CO_2)_9$	hexaaquazinc sodium triuranyl nonaacetate
95. $MgNH_4PO_4 \cdot 6H_2O$	ammonium magnesium phosphate—water (1/6) or ammonium magnesium phosphate hexahydrate
96. $NaCl \cdot NaF \cdot 2Na_2SO_4$	hexasodium chloride fluoride bis(sulfate), empirical formula $Na_6ClF(SO_4)_2$
97. $Ca_5F(PO_4)_3$	pentacalcium fluoride tris(phosphate)

The recommended methods of naming hydrates and double salts should be evident from these examples.

The names of groups that can be regarded as substituents in organic compounds or as ligands on metals are often the same as the names of the corresponding radicals. Names of radicals (and of the related substituent groups) are generally derived from parent hydride names by modifying their names with the suffix -yl, according to the rules of substitutive nomenclature. Sometimes contractions are used.

Examples

98. $SiH_3^{\cdot}$	SiH_3^{-}	silyl
99. $SnCl_3^{\cdot}$	$SnCl_3^{-}$	trichlorostannyl
100. $BH_2^{\cdot}$	BH_2^{-}	boryl
101. $CH_3^{\cdot}$	CH_3^{-}	methyl

In inorganic chemistry, certain substituent groups retain trivial names that are still in general use.

Examples

102. OH	hydroxyl
103. CO	carbonyl
104. SO_2	sulfonyl
105. CrO_2	chromyl

A complete list is given in Table 4.1. These names may be used in inorganic functional class names.

Examples

106. $COCl_2$	carbonyl chloride
107. NOCl	nitrosyl chloride
108. SO_2NH	sulfuryl imide
109. IO_2F	iodyl fluoride

Element substituent group names are formed by adding the suffix -io to the stem of the name (compare anion name formation).

Examples

110. Cl	chlorio (chloro if in oxidation state $-I$)
111. Na	sodio
112. ClHg	chloromercurio
113. $(OC)_4Co$	tetracarbonylcobaltio
114. F_5S	pentafluorosulfurio

The full list of such names is given in Table 4.3.

4.3   MORE COMPLEX NOMENCLATURE SYSTEMS

When it is required to convey more information than is implied by a simple compositional name, other approaches to name construction are adopted. The structure of the compound under consideration generally dictates the name adopted, even though a compound may be named correctly in more ways than one. For molecular compounds, substitutive nomenclature, originally developed for naming organic compounds and the oldest systematic nomenclature still in use, is generally used.

This system, which relies upon the concept of a parent compound from which a series of products may be derived in a formal fashion by replacement (or, otherwise,

Table 4.3 Names of elements as substituent groups.*

Element name	Radical name	Element name	Radical name
Actinium	Actinio	Mercury	Mercurio
Aluminium	Aluminio	Molybdenum	Molybdenio
Americium	Americio	Neodymium	Neodymio
Antimony	Antimonio	Neon	Neonio
Argon	Argonio	Neptunium	Neptunio
Arsenic	Arsenio	Nickel	Nickelio
Astatine	Astatio	Niobium	Niobio
Barium	Bario	Nitrogen	–
Berkelium	Berkelio	Nobelium	Nobelio
Beryllium	Beryllio	Osmium	Osmio
Bismuth	Bismuthio	Oxygen	–
Boron	Borio	Palladium	Palladio
Bromine	Bromio	Phosphorus	Phosphorio
Cadmium	Cadmio	Platinum	Platinio
Caesium	Caesio	Plutonium	Plutonio
Calcium	Calcio	Polonium	Polonio
Californium	Californio	Potassium	Potassio (kalio)
Carbon	–	Praseodymium	Praseodymio
Cerium	Cerio	Promethium	Promethio
Chlorine	Chlorio	Protactinium	Protactinio
Chromium	Chromio	Radium	Radio
Cobalt	Cobaltio	Radon	Radonio
Copper (cuprum)	Cuprio	Rhenium	Rhenio
Curium	Curio	Rhodium	Rhodio
Deuterium	Deuterio	Rubidium	Rubidio
Dysprosium	Dysprosio	Ruthenium	Ruthenio
Einsteinium	Einsteinio	Samarium	Samario
Erbium	Erbio	Scandium	Scandio
Europium	Europio	Selenium	Selenio
Fermium	Fermio	Silicon	Silicio
Fluorine	Fluorio	Silver (argentum)	Argentio
Francium	Francio	Sodium	Sodio (natrio)
Gadolinium	Gadolinio	Strontium	Strontio
Gallium	Gallio	Sulfur	Sulfurio
Germanium	Germanio	Tantalum	Tantalio
Gold (aurum)	Aurio	Technetium	Technetio
Hafnium	Hafnio	Tellurium	Tellurio
Helium	Helio	Terbium	Terbio
Holmium	Holmio	Thallium	Thallio
Hydrogen	–	Thorium	Thorio
Indium	Indio	Thulium	Thulio
Iodine	Iodio	Tin (stannum)	Stannio
Iridium	Iridio	Titanium	Titanio
Iron (ferrum)	Ferrio	Tritium	Tritio
Krypton	Kryptonio	Tungsten (wolfram)	Tungstenio (wolframio)
Lanthanum	Lanthanio	Uranium	Uranio
Lawrencium	Lawrencio	Vanadium	Vanadio
Lead (plumbum)	Plumbio	Xenon	Xenonio
Lithium	Lithio	Ytterbium	Ytterbio
Lutetium	Lutetio	Yttrium	Yttrio
Magnesium	Magnesio	Zinc	Zincio
Manganese	Manganio	Zirconium	Zirconio
Mendelevium	Mendelevio		

* These names are used in organic substitutive nomenclature for situations in which the substituent group is joined to the parent skeleton by a single element–carbon bond.

substitution), generally of a hydrogen atom by another atom or group (substituent), is not easily applied to coordination compounds, for which an additive system has been developed. This additive system arose from the concepts of complex formation devised by Werner around the beginning of the twentieth century. It can be used also for compounds formally derived by replacing a skeletal carbon atom of a parent hydride by a heteroatom such as silicon or even a metal atom. This usage will be described below.

Polymeric compounds (macromolecules) do not fall easily into either of these categories, and for them a subsystem of macromolecular nomenclature has been developed. A brief introduction to macromolecular nomenclature is presented in Chapter 6. Non-stoichiometric compounds also are clearly difficult to name within the constraints of a description which generally implies localised electron-pair bonds or specific atom–atom interactions. For these, further systems of nomenclature are in the process of development. Finally, oxoacids and inorganic rings and chains have their own nomenclature variants.

This is not an exhaustive list, but it illustrates the fact that the choice of an appropriate naming method is a function of the substance to be named, and that there may be more than one way of deriving a correct name.

4.4   COORDINATION NOMENCLATURE, AN ADDITIVE NOMENCLATURE

4.4.1   Introduction

Coordination nomenclature was developed using the same concepts that were developed to categorise coordination compounds. It was recognised that, although metals exhibit what were termed 'primary valencies' in compounds such as $NiCl_2$, $Fe_2(SO_4)_3$ and $PtCl_2$, addition compounds could be formed in which 'secondary valencies' were also exhibited. Examples of such compounds include $NiCl_2 \cdot 4H_2O$ and $PtCl_2 \cdot 2KCl$. The exercise of this secondary valence was essentially recognised as an expression of metal–ligand coordination. This gave rise to the Werner theory of coordination compounds, and coordination nomenclature is most easily, although not exclusively, applied to coordination compounds.

Coordination nomenclature relies on the identification of a coordination entity: a central atom (usually a metal) surrounded by a set of ligands. In the original electronic formulations, the bonding between the metal atom and the ligand involved the sharing of a lone pair of electrons of the ligand (donor) with the metal (acceptor). Clearly, terms such as 'free ligand' and 'to ligand' are etymologically and logically unsound, despite their adoption in some less enlightened circles. Typical formulae of coordination entities are $[CoCl_4]^{2-}$, $[Co_2(CO)_8]$, $[NiCl_2(PEt_3)_2]$ and $[VCl_3(NCMe)_3]$. Note the use of square brackets to define the entity, whether charged or not. This usage differs from that in organic nomenclature systems. A polymeric material is designated $\{MCl_x\}_n$ or, if desired, $[\{MCl_x\}_n]$, but never $[MCl_x]_n$. Similarly, the empirical formula $PtCl_2(PEt_3)$ corresponds to the dinuclear species $[\{PtCl_2(PEt_3)\}_2]$, which should never be written $[PtCl_2(PEt_3)]_2$.

Clearly, there can be name duplication when coordination nomenclature principles are applied to systems not normally regarded as coordination compounds. Thus,

[SiCl$_4$(C$_5$H$_5$N)$_2$] is a coordination compound. Formally, one might represent SiCl$_4$ itself as Si^{IV} (or Si^{4+}) with four chloride ligands. To insist on a formulation [SiCl$_4$] would be mere pedantry. Nevertheless, a coordination name (see below) might be appropriate in some circumstances.

4.4.2 Definitions

A coordination entity is composed of a central atom or atoms to which are attached other atoms or groups of atoms, which are termed ligands. A central atom occupies a central position within the coordination entity. The ligands attached to a central atom define a coordination polyhedron. Each ligand is assumed to be at the vertex of an appropriate polyhedron. The usual polyhedra are shown in Table 3.3 and they are also listed in Table 4.4. Note that these are adequate to describe most simple coordination compounds, but that real molecules do not always fall into these simple categories. In the presentation of a coordination polyhedron graphically, the lines defining the polyhedron edges are not indicative of bonds.

The most common 'polyhedra' encountered in simple coordination chemistry are the square, the tetrahedron and the octahedron.

At this level of approximation, each bond between the metal and a ligand is considered to be a two-electron sigma bond. For the purposes of electron counting (as in the inert-gas rule), this is normally adequate. The number of such bonds is the coordination number. Each single, simple ligand contributes unity to this number. Thus, a square planar coordination implies a coordination number of 4. An octahedron implies a value of 6. The typical values are also included in Table 4.4.

However, many ligands do not behave as donors of a single electron pair. Some ligands donate two or more electron pairs to the same central atom from different donor atoms. Such ligands are said to be chelating ligands, and they form chelate rings, closed by the central atom. The phenomenon is termed chelation.

Example

Table 4.4 List of polyhedral symbols.*

Coordination polyhedron	Coordination number	Polyhedral symbol
Linear	2	L-2
Angular	2	A-2
Trigonal plane	3	TP-3
Trigonal pyramid	3	TPY-3
Tetrahedron	4	T-4
Square plane	4	SP-4
Square pyramid	4	SPY-4
Trigonal bipyramid	5	TBPY-5
Square pyramid	5	SPY-5
Octahedron	6	OC-6
Trigonal prism	6	TPR-6
Pentagonal bipyramid	7	PBPY-7
Octahedron, face monocapped	7	OCF-7
Trigonal prism, square face monocapped	7	TPRS-7
Cube	8	CU-8
Square antiprism	8	SAPR-8
Dodecahedron	8	DD-8
Hexagonal bipyramid	8	HBPY-8
Octahedron, trans-bicapped	8	OCT-8
Trigonal prism, triangular face bicapped	8	TPRT-8
Trigonal prism, square face bicapped	8	TPRS-8
Trigonal prism, square face tricapped	9	TPRS-9
Heptagonal bipyramid	9	HBPY-9

* Strictly, not all the geometries can be represented by polyhedra.

Sometimes bonds that involve a neutral ligand donating an electron pair are represented by an arrow as from $H_2NCH_2CH_2NH_2$ above, and bonds that can formally be regarded as involving an electron from each partner are shown, as usual, by a line. These formalisms are arbitrary and are not recommended.

The number of electron pairs donated by a single ligand to a specific central atom is termed the denticity. Ligands that donate one pair are monodentate, those that donate two are didentate, those that donate three are tridentate, and so on.

Example

tetradentate coordination

Sometimes ligands with two or more potential donor sites bond to two (or more) different central atoms rather than to one, forming a bridge between central atoms. It may not be necessary for the ligand in such a system to be like ethane-1,2-diamine, with two distinct potential donor atoms. A donor atom with two or more pairs of non-bonding electrons in its valence shell can also donate them to different central atoms. Such ligands, of whatever type, are called bridging ligands. They bond to two or more central atoms simultaneously. The number of central atoms in a single coordination entity is denoted by the nuclearity: mononuclear, dinuclear, trinuclear, etc. Atoms that can bridge include S, O and Cl.

Examples

3.

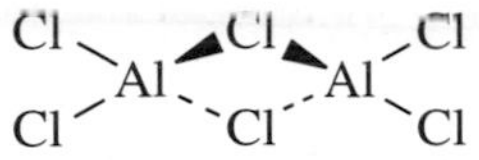

Al_2Cl_6, a dinuclear complex with bridging chloride ions.

4.

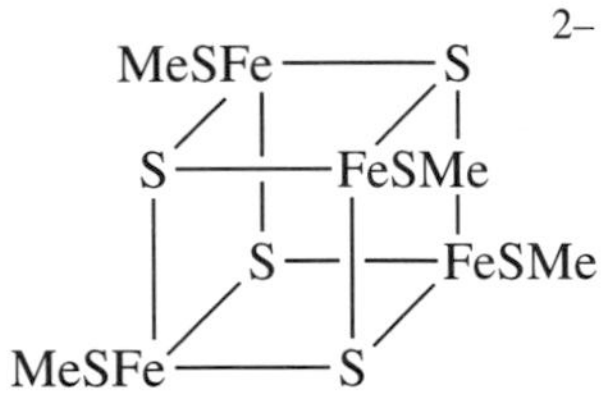

$[Fe_4S_4(SMe)_4]^{2-}$, a tetranuclear complex with bridging sulfide ions.

5. $[(NH_3)_5CoNH_2CH_2CH_2NH_2Co(NH_3)_5]^{6+}$
This contains bis(monodentate) bridging ethane-1,2-diamine.

The original concepts of metal–ligand bonding were essentially related to the dative covalent bond; the development of organometallic chemistry has revealed a further way in which ligands can supply more than one electron pair to a central atom. This is exemplified by the classical cases of bis(benzene)chromium and bis(cyclopentadienyl)iron, trivial name ferrocene. These molecules are characterised by the bonding of a formally unsaturated system (in the organic chemistry sense, but expanded to include aromatic systems) to a central atom, usually a metal atom.

Examples

6. 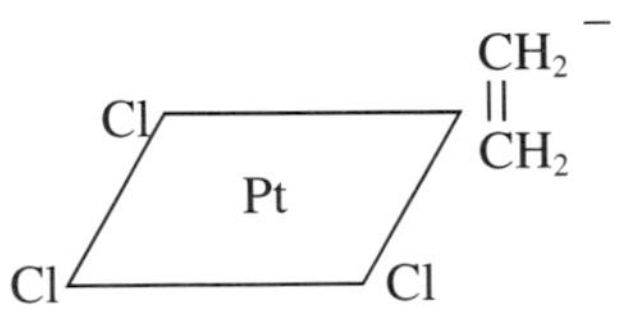   7. 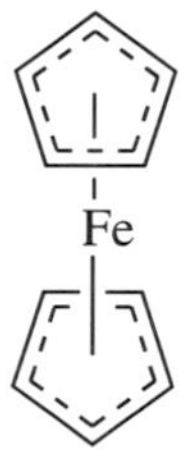 Fe   8. 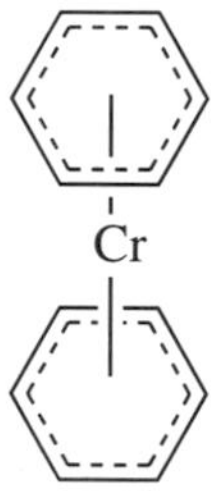 Cr

In the simpler cases, such as ethylene, the π-electron pair can be donated to the metal just like a lone pair on, say, the nitrogen of ammonia. This results in a contribution of unity to the coordination number, but two carbon atoms are bound to the central atom. The hapticity of the ethylene is defined as two and is denoted formally by the symbol η^2. In general, the hapticity of a ligand is the number of ligating atoms, n, in the ligand that bind to the metal, and is represented by the symbol η^n.

In the bis(benzene)chromium molecule, each double bond of each benzene molecule may be considered as donating two electrons. The benzene molecules are thus η^6, and the coordination number of the chromium is also 6. Molecules with odd numbers of carbon atoms can also be involved, although in such a case a formal charge must be assigned to the hydrocarbon molecule. Thus in bis(cyclopentadienyl)iron, each cyclopentadienyl is considered to contribute six electrons to the iron, i.e. it is formally regarded as cyclopentadienide, $C_5H_5^-$. Each ligand is pentahapto or η^5, but each supplies six electrons.

One might care to define the coordination number of the iron atom in this case as 6, but the concept is really losing some of its clarity in such instances. The important factor is six electrons. A further point to note is that, although the organic ligands are named as radicals, they are formally treated as anions. This is generally true of organic groups in coordination nomenclature, and includes methyl, ethyl, allyl and phenyl. This discussion implies that the hapto symbol, η, is not strictly related to electron pairs. It defines the manner in which the ligand binds to the central atom. Thus, even η^1 is allowed, although in terms of its original IUPAC definition that required the donating atoms of the ligand to be contiguous, this is a nonsense.

One final concept needs to be mentioned. In many compounds, two central atoms can be bridged by a direct bond between them, without any bridging ligand. This bond occupies a position in the coordination shell without using a lone pair. Essentially, the two central atoms share pairs of electrons, one electron coming from each central atom in most cases. This is reflected in the corresponding oxidation states.

Examples

9. $[Br_4Re\text{-}ReBr_4]^{2+}$
10. $[(CO)_5Re\text{-}Co(CO)_4]$

Nomenclature practice does not define the multiplicity of such bonds, be they single, double, triple or even quadruple. It adopts a device to indicate that a bond exists between the metal atoms. This will be presented below.

4.4.3   **Mononuclear coordination compounds**

4.4.3.1   *Formulae.* The central atom is listed first. The formally anionic ligands appear next, listed in alphabetical order of the first symbols of their individual formulae. The neutral ligands follow, also in alphabetical order. Polydentate ligands are included in alphabetical order, the formula to be presented as discussed in Chapter 3. The formula for the entire coordination entity, whether charged or not, is enclosed in square brackets. For coordination formulae, the nesting order of enclosing marks is as given on p. 13. The charge on an ion is indicated in the usual way by use of a right superscript. Oxidation states of particular atoms are indicated by an appropriate roman numeral as a right superscript to the symbol of the atom in question, and not in parentheses on the line. In the formula of a salt containing coordination entities, cation always precedes anion, no charges are indicated and there is no space between the formulae for cation and anion.

Examples

1. $[Co(NH_3)_6]Cl$ 2. $[PtCl_4]^{2-}$

3. $[CoCl(NH_3)_5]Cl$
4. $Na[PtBrCl(NO_2)(NH_3)]$
5. $[CaCl_2\{OC(NH_2)_2\}_2]$

6. $[Cr^{III}(NCS)_4(NH_3)_2]^-$
7. $[Fe^{-II}(CO)_4]^{2-}$

The precise form of a formula should be dictated by the needs of the user. For example, it is generally recommended that a ligand formula within a coordination formula be written so that the donor atom comes first, e.g. $[TiCl_3(NCMe)_3]$, but this is not mandatory and should not affect the recommended order of ligand citation. It may also be impossible to put all the donor atoms first, e.g. where two donors are present in a chelate complex: $[Co(NH_2CH_2CH_2NH_2)_3]^{3+}$. Whether the ethane-1,2-diamine is displayed as shown, or simply aggregated as $[Co(C_2H_8N_2)_3]^{3+}$, is a matter of choice. Certainly there is a conflict between this last form and the suggestion that the donor atoms be written first. The aim should always be clarity, at the expense of rigid adherence to recommendations.

It is often inconvenient to represent all the ligand formulae in detail. Abbreviations are often used and are indeed encouraged, with certain provisos. These are: the abbreviations should all be written in lower case (with minor exceptions, such as Me, Et and Ph) and preferably not more than four letters; with certain exceptions of wide currency, abbreviations should be defined in a text when they first appear; in a formula, the abbreviation should be enclosed in parentheses, and its place in the citation sequence should be determined by its formula, as discussed above; and particular attention should be paid to the loss of hydrons from a ligand precursor.

This last proviso is exemplified as follows. Ethylenediaminetetraacetic acid should be rendered H_4edta. The ions derived from it, which are often ligands in coordination entities, are then $(H_3edta)^-$, $(H_2edta)^{2-}$, $(Hedta)^{3-}$ and $(edta)^{4-}$. This avoids monstrosities such as $edta-H_2$ and $edtaH_{-2}$ which arise if the parent acid is represented as edta. A list of recommended abbreviations is presented in Table 4.5.

4.4.3.2   *Names.* The addition of ligands to a central atom is paralleled in name construction. The names of the ligands are added to that of the central atom. The ligands are listed in alphabetical order regardless of ligand type. Numerical prefixes are ignored in this ordering procedure, unless they are part of the ligand name. Charge number and oxidation number are used as necessary in the usual way.

Of the two kinds of numerical prefix (see Table 4.2), the simple di-, tri-, tetra-, etc. are generally recommended. The prefixes bis-, tris-, tetrakis-, etc. are to be used only with more complex expressions and to avoid ambiguity. They normally require parentheses around the name they qualify. The nesting order of enclosing marks is as cited on p. 13. There is normally no elision in instances such as tetraammine and the two adjacent letters 'a' are pronounced separately.

The names of ligands recommended for general purposes are given in Table 4.6. The names for anionic ligands end in -o. If the anion name ends in -ite, -ate or -ide, the ligand name is changed to -ito, -ato or -ido. The halogenido names are, by custom, abbreviated to halo. Note that hydrogen as a ligand is always regarded as anionic, with the name hydride. The names of neutral and cationic ligands are never modified. Water and ammonia molecules as ligands take the names aqua and ammine, respectively. Parentheses are always placed around ligand names, which themselves contain multiplicative prefixes, and are also used to ensure clarity, but

Table 4.5 Representation of ligand names by abbreviation.*

Abbreviation	Common name	Systematic name
Diketones		
Hacac	acetylacetone	2,4-pentanedione
Hhfa	hexafluoroacetylacetone	1,1,1,5,5,5-hexafluoropentane-2,4-dione
Hba	benzoylacetone	1-phenylbutane-1,3-dione
Hfod	1,1,1,2,2,3,3-heptafluoro-7,7-dimethyl-4,6-octanedione	6,6,7,7,8,8,8-heptafluoro-2,2-dimethyl octane-3,5-dione
Hfta	trifluoroacetylacetone	1,1,1-trifluoropentane-2,4-dione
Hdbm	dibenzoylmethane	1,3-diphenylpropane-1,3-dione
Hdpm	dipivaloylmethane	2,2,6,6-tetramethylheptane-3,5-dione
Amino alcohols		
Hea	ethanolamine	2-aminoethanol
H$_3$tea	triethanolamine	2,2′,2′′-nitrilotriethanol
H$_2$dea	diethanolamine	2,2′-iminodiethanol
Hydrocarbons		
cod	cyclooctadiene	cycloocta-1,5-diene
cot	cyclooctatetraene	cycloocta-1,3,5,7-tetraene
Cp	cyclopentadienyl	cyclopentadienyl
Cy	cyclohexyl	cyclohexyl
Ac	acetyl	acetyl
Bu	butyl	butyl
Bzl	benzyl	benzyl
Et	ethyl	ethyl
Me	methyl	methyl
nbd	norbornadiene	bicyclo[2.2.1]hepta-2,5-diene
Ph	phenyl	phenyl
Pr	propyl	propyl
Heterocycles		
py	pyridine	pyridine
thf	tetrahydrofuran	tetrahydrofuran
Hpz	pyrazole	1*H*-pyrazole
Him	imidazole	1*H*-imidazole
terpy	2,2′,2′′-terpyridine	2,2′:6′,2′′-terpyridine
picoline	α-picoline	2-methylpyridine
Hbpz$_4$	hydrogen tetra(1-pyrazolyl)borate(1−)	hydrogen tetrakis(1*H*-pyrazolato-*N*)borate(1−)
isn	isonicotinamide	4-pyridinecarboxamide
nia	nicotinamide	3-pyridinecarboxamide
pip	piperidine	piperidine
lut	lutidine	2,6-dimethylpyridine
Hbim	benzimidazole	1*H*-benzimidazole
Chelating and other ligands		
H$_4$edta	ethylenediaminetetraacetic acid	(ethane-1,2-diyldinitrilo)tetraacetic acid
H$_5$dtpa	*N,N,N′,N′′,N′′*-diethylenetriaminepentaacetic acid	[[(carboxymethyl)imino]bis(ethane-1,2-diylnitrilo)]tetraacetic acid
H$_3$nta	nitrilotriacetic acid	
H$_4$cdta	*trans*-1,2-cyclohexanediaminetetraacetic acid	*trans*-(cyclohexane-1,2-diyldinitrilo)tetraacetic acid
H$_2$ida	iminodiacetic acid	iminodiacetic acid
dien	diethylenetriamine	*N*-(2-aminoethyl)ethane-1,2-diamine
en	ethylenediamine	ethane-1,2-diamine
pn	propylenediamine	propane-1,2-diamine
tmen	*N,N,N′,N′*-tetramethylethylenediamine	*N,N,N′,N′*-tetramethylethane-1,2-diamine
tn	trimethylenediamine	propane-1,3-diamine

Continued on p. 58.

Table 4.5 (*Continued.*)

Abbreviation	Common name	Systematic name
tren	tris(2-aminoethyl)amine	N,N-bis(2-aminoethyl)ethane-1,2-diamine
trien	triethylenetetramine	N,N'-bis(2-aminoethyl)ethane-1,2-diamine
chxn	1,2-diaminocyclohexane	cyclohexane-1,2-diamine
hmta	hexamethylenetetramine	1,3,5,7-tetraazatricyclo[3.3.1.1^{3,7}]decane
Hthsc	thiosemicarbazide	hydrazinecarbothioamide
depe	1,2-bis(diethylphosphino)ethane	ethane-1,2-diylbis(diethylphosphine)
diars	*o*-phenylenebis(dimethylarsine)	1,2-phenylenebis(dimethylarsine)
dppe	1,2-bis(diphenylphosphino)ethane	ethane-1,2-diylbis(diphenylphosphine)
diop	2,3-*O*-isopropylidene-2,3-dihydroxy-1,4-bis(diphenylphosphino)butane	3,4-bis[(diphenylphosphinyl)methyl]-2,2-dimethyl-1,3-dioxolane
triphos		[2-[(diphenylphosphino)methyl]-2-methyl propane-1,3-diyl]bis(diphenylphosphine)
hmpa	hexamethylphosphoric triamide	hexamethylphosphoric triamide
bpy	2,2'-bipyridine	2,2'-bipyridine
H$_2$dmg	dimethylglyoxime	2,3-butanedione dioxime
dmso	dimethyl sulfoxide	sulfinyldimethane
phen	1,10-phenanthroline	1,10-phenanthroline
tu	thiourea	thiourea
Hbig	biguanide	imidodicarbonimidic diamide
HEt$_2$dtc	diethyldithiocarbamic acid	diethylcarbamodithioic acid
H$_2$mnt	maleonitriledithiol	2,3-dimercaptobut-2-enedinitrile
tcne	tetracyanoethylene	ethenetetracarbonitrile
tcnq	tetracyanoquinodimethan	2,2'-(cyclohexa-2,5-diene-1,4-diylidene)bis(1,3-propanedinitrile)
dabco	triethylenediamine	1,4-diazabicyclo[2.2.2]octane
2,3,2-tet	1,4,8,11-tetraazaundecane	N,N'-bis(2-aminoethyl)propane-1,3-diamine
3,3,3-tet	1,5,9,13-tetraazatridecane	N,N'-bis(3-aminopropyl)propane-1,3-diamine
ur	urea	urea
dmf	dimethylformamide	N,N-dimethylformamide

Schiff base

H$_2$salen	bis(salicylidene)ethylenediamine	2,2'-[ethane-1,2-diylbis(nitrilomethylidyne)]diphenol
H$_2$acacen	bis(acetylacetone)ethylenediamine	4,4'-(ethane-1,2-diyldinitrilo)bis(pentan-2-one)
H$_2$salgly	salicylideneglycine	N-[(2-hydroxyphenyl)methylene]glycine
H$_2$saltn	bis(salicylidene)-1,3-diaminopropane	2,2'-[propane-1,3-diylbis(nitrilomethylidyne)]diphenol
H$_2$saldien	bis(salicylidene)diethylenetriamine	2,2'-[iminobis(ethane-1,2-diylnitrilomethylidyne)]diphenol
H$_2$tsalen	bis(2-mercaptobenzylidene)ethylenediamine	2,2'-[ethane-1,2-diylbis(nitrilomethylidyne)]dibenzenethiol

Macrocycles

18-crown-6	1,4,7,10,13,16-hexaoxacyclooctadecane	1,4,7,10,13,16-hexaoxacyclooctadecane
benzo-15-crown-5	2,3-benzo-1,4,7,10,13-pentaoxacyclopentadec-2-ene	2,3,5,6,8,9,11,12-octahydro-1,4,7,10,13-benzopentaoxacyclopentadecine
cryptand 222	4,7,13,16,21,24-hexaoxa-1,10-diazabicyclo[8.8.8]hexacosane	4,7,13,16,21,24-hexaoxa-1,10-diazabicyclo[8.8.8]hexacosane
cryptand 211	4,7,13,18-tetraoxa-1,10-diazabicyclo[8.5.5]icosane	4,7,13,18-tetraoxa-1,10-diazabicyclo[8.5.5]icosane
[12]aneS$_4$	1,4,7,10-tetrathiacyclododecane	1,4,7,10-tetrathiacyclododecane
H$_2$pc	phthalocyanine	phthalocyanine
H$_2$tpp	tetraphenylporphyrin	5,10,15,20-tetraphenylporphyrin
H$_2$oep	octaethylporphyrin	2,3,7,8,12,13,17,18-octaethylporphyrin
ppIX	protoporphyrin IX	3,7,12,17-tetramethyl-8,13-divinylporphyrin-2,18-dipropanoic acid

Continued.

Table 4.5 (*Continued.*)

Abbreviation	Common name	Systematic name
[18]aneP$_4$O$_2$	1,10-dioxa-4,7,13,16-tetraphosphacyclooctadecane	1,10-dioxa-4,7,13,16-tetraphosphacyclooctadecane
[14]aneN$_4$	1,4,8,11-tetraazacyclotetradecane	1,4,8,11-tetraazacyclotetradecane
[14]1,3-dieneN$_4$	1,4,8,11-tetraazacyclotetradeca-1,3-diene	1,4,8,11-tetraazacyclotetradeca-1,3-diene
Me$_4$[14]-aneN$_4$	2,3,9,10-tetramethyl-1,4,8,11-tetraazacyclotetradecane	2,3,9,10-tetramethyl-1,4,8,11-tetraazacyclotetradecane
cyclam		1,4,8,11-tetraazacyclotetradecane

* The following practices should be followed in the use of abbreviations. It should be assumed that the reader will not be familiar with the abbreviations. Consequently, all text should explain the abbreviations it uses. The abbreviations in this table are widely used, and it is hoped that they will become standard. The commonly accepted abbreviations for organic groups (Me, methyl; Et, ethyl; Ph, phenyl; etc.) should not be used with any other meanings. The most useful abbreviations are those that readily suggest the ligand in question, either because they are obviously derived from the ligand name or because they are systematically related to structure. The sequential positions of ligand abbreviations in formulae should be in accordance with Section 4.4.3.1. Lower case letters are used for all abbreviations, except for those of certain hydrocarbon radicals. In formulae, the ligand abbreviation should be in parentheses, as in [Co(en)$_3$]$^{3+}$. Those hydrogen atoms that can be replaced by the metal atom are shown in the abbreviation by the symbol H. Thus, the molecule Hacac forms an anionic ligand that is abbreviated acac.

aqua, ammine, carbonyl (CO) and nitrosyl (NO) do not require them.

The names of all cationic and neutral entities end in the name of the element, together with the charge (if appropriate) or the oxidation state (if desired). The names of complex anions require modification, and this is achieved by adding the termination -ate. All these recommendations are illustrated in the following examples.

Examples

1. Dichloro(**d**iphenylphosphine)(thiourea)platinum(II)
2. K$_4$[Fe(CN)$_6$]   potassium hexacyanoferrate(II)
   potassium hexacyanoferrate(4−)
   tetrapotassium hexacyanoferrate
3. [Co(NH$_3$)$_6$]Cl$_3$ hexaamminecobalt(III) chloride
4. [CoCl(NH$_3$)$_5$]Cl$_2$ pentaamminechlorocobalt(2+) chloride
5. [CoCl(NO$_2$)(NH$_3$)$_4$]Cl   tetraamminechloronitrito-*N*-cobalt(III) chloride
6. [PtCl(NH$_2$CH$_3$)(NH$_3$)$_2$]Cl   diamminechloro(methylamine)platinum(II) chloride
7. [CuCl$_2${OC(NH$_2$)$_2$}$_2$]   dichlorobis(urea)copper(II)
8. K$_2$[PdCl$_4$]   potassium tetrachloropalladate(II)
9. K$_2$[OsCl$_5$N]   potassium pentachloronitridoosmate(2−)
10. Na[PtBrCl(NO$_2$)(NH$_3$)]   sodium amminebromochloronitrito-*N*-platinate(1−)
11. [Fe(CNCH$_3$)$_6$]Br$_2$ hexakis(methyl isocyanide)iron(II) bromide
12. [Ru(HSO$_3$)$_2$(NH$_3$)$_4$]   tetraamminebis(hydrogensulfito)ruthenium(II)
13. [Co(H$_2$O)$_2$(NH$_3$)$_4$]Cl$_3$ tetraamminediaquacobalt(III) chloride
14. [PtCl$_2$(C$_5$H$_5$N)(NH$_3$)]   amminedichloro(pyridine)platinum(II)
15. Ba[BrF$_4$]$_2$ barium tetrafluorobromate(III)
16. K[CrF$_4$O]   potassium tetrafluorooxochromate(V)
17. [Ni(H$_2$O)$_2$(NH$_3$)$_4$]SO$_4$ tetraamminediaquanickel(II) sulfate

Table 4.6 Names of ligands.*

Formula	Systematic name	Alternative ligand name
Ligands based on Group 15 elements		
N_2	(dinitrogen)	
P_4	(tetraphosphorus)	
As_4	(tetraarsenic)	
N^{3-}	nitrido	
P^{3-}	phosphido	
As^{3-}	arsenido	
$(N_2)^{2-}$	[dinitrido(2–)]	
$(N_2)^{4-}$	[dinitrido(4–)]	[hydrazido(4–)]
$(N_3)^-$	(trinitrido)	azido†
$(P_2)^{2-}$	[diphosphido(2–)]	
$(CN)^-$	cyano	
$(NCO)^-$	(cyanato)	
$(NCS)^-$	(thiocyanato)	
$(NCSe)^-$	(selenocyanato)	
$(NCN)^{2-}$	[carbodiimidato(2–)]	
NF_3	(trifluoroazane)	(nitrogen trifluoride)
NH_3	(azane)	ammine†,‡
PH_3	(phosphane)	(phosphine)
AsH_3	(arsane)	(arsine)
SbH_3	(stibane)	(stibine)
$(NH)^{2-}$	azanediido	imido†,‡
$(NH_2)^-$	azanido	amido†,‡
$(PH)^{2-}$	phosphanediido	phosphinidene‡
$(PH_2)^-$	phosphanido	phosphino‡
$(SbH)^{2-}$	stibanediido	stibylene‡
$(SbH_2)^-$	stibanido	stibino‡
$(AsH)^{2-}$	arsanediido	arsinidene‡
$(AsH_2)^-$	arsanido	arsino‡
$(FN)^{2-}$	(fluoroazanediido)	(fluorimido)
$(ClHN)^-$	(chloroazanido)	(chloramido)
$(Cl_2N)^-$	(dichloroazanido)	(dichloramido)
$(FP)^{2-}$	(fluorophosphanediido)	
$(F_2P)^-$	(difluorophosphanido)	(difluorophosphido)
		(phosphonous difluoridato)‡
CH_3NH_2	(methanamine)	(methylamine)§
$(CH_3)_2NH$	(*N*-methylmethanamine)	(dimethylamine)§
$(CH_3)_3N$	(*N,N*-dimethylmethanamine)	(trimethylamine)§
CH_3PH_2	(methylphosphane)	(methylphosphine)
$(CH_3)_2PH$	(dimethylphosphane)	(dimethylphosphine)
$(CH_3)_3P$	(trimethylphosphane)	(trimethylphosphine)
$(CH_3N)^{2-}$	[methanaminato(2–)]	(methylimido)
$(CH_3NH)^-$	[methanaminato(1–)]	(methylamido)
$[(CH_3)_2N]^-$	(*N*-methylmethanaminato)	(dimethylamido)
$[(CH_3)_2P]^-$	(dimethylphosphanido)	(dimethylphosphino)
$(CH_3P)^{2-}$	(methylphosphanediido)	(methylphosphinidene)‡
$(CH_3PH)^-$	(methylphosphanido)	(methylphosphino)
$HN{=}NH$	(diazene)	(diimide), (diimine)
H_2NNH_2	(diazane)	(hydrazine)
HN_3	(hydrogen trinitride)	(hydrogen azide)†
$(HN{=}N)^-$	(diazenido)	(diiminido)
$(HNN)^{3-}$	(diazanetriido)	[hydrazido(3–)]
$(H_2NN)^{2-}$	(diazane-1,1-diido)	[hydrazido(2–)-*N,N*]
$(HN{-}NH)^{2-}$	(diazane-1,2-diido)	[hydrazido(2–)-*N,N'*]
$(H_2N{-}NH)^-$	(diazanido)	(hydrazido)
$HP{=}PH$	(diphosphene)	
$H_2P{-}PH_2$	(diphosphane)	

Continued.

Table 4.6 (*Continued.*)

Formula	Systematic name	Alternative ligand name
$(HP{=}P)^-$	(diphosphenido)	
$(H_2P{-}P)^{2-}$	(diphosphane-1,1-diido)	
$(HP{-}PH)^{2-}$	(diphosphane-1,2-diido)	
$(H_2PPH)^-$	(diphosphanido)	
$HAs{=}AsH$	(diarsene)	
H_2AsAsH_2	(diarsane)	
$(HAsAs)^{3-}$	(diarsanetriido)	
$(H_2AsAs)^{2-}$	(diarsane-1,1-diido)	
$(CH_3AsH)^-$	(methylarsanido)	(methylarsino)‡
$(CH_3As)^{2-}$	(methylarsanediido)	(methylarsinidene)‡
H_2NOH	(hydroxyazane)	(hydroxylamine)
$(HNOH)^-$	(hydroxylaminato-κN)	(hydroxylamido)
$(H_2NO)^-$	(hydroxylaminato-κO)	(hydroxylamido)
$(HNO)^{2-}$	[hydroxylaminato(2–)]	(hydroxylimido)
$(PO_3)^{3-}$	[trioxophosphato(3–)]	[phosphito(3–)]
$(HPO_2)^{2-}$	[hydridodioxophosphato(2–)]	[phosphonito(2–)]
$(H_2PO)^-$	[dihydridooxophosphato(1–)]	(phosphonito)
$(AsO_3)^{3-}$	[trioxoarsenato(3–)]	[arsenito(3–)]
$(HAsO_2)^{2-}$	[hydridodioxoarsenato(2–)]	[arsenito(2–)]
$(H_2AsO)^-$	[dihydridooxoarsenato(1–)]	(arsinito)
$(PO_4)^{3-}$	[tetraoxophosphato(3–)]	[phosphato(3–)]
$(HPO_3)^{2-}$	[hydridotrioxophosphato(2–)]	[phosphonato(2–)]
$(H_2PO_2)^-$	[dihydridodioxophosphato(1–)]	(phosphinato)
$(AsO_4)^{3-}$	[tetraoxoarsenato(3–)]	[arsenato(3–)]
$(HAsO_3)^{2-}$	[hydridotrioxoarsenato(2–)]	[arsonato(2–)]
$(H_2AsO_2)^-$	[dihydridodioxoarsenato(1–)]	arsinato
$(P_2O_7)^{4-}$	[μ-oxo-hexaoxodiphosphato(4–)]	[diphosphato(4–)]
$(C_6H_5N_2)^-$	(phenyldiazenido)	(phenylazo)§
$(NO_2)^-$	[dioxonitrato(1–)]	nitrito-O
	[dioxonitrato(1–)]	nitrito-N, nitro
$(NO_3)^-$	[trioxonitrato(1–)]	nitrato
NO	(nitrogen monoxide)	nitrosyl
NS	(nitrogen monosulfide)	(thionitrosyl)
N_2O	(dinitrogen oxide)	
$(N_2O_2)^{2-}$	[dioxodinitrato($N{-}N$)(2–)]	hyponitrito
Ligands based on chalcogen elements (elements of Group 16)		
O_2	(dioxygen)	oxygen
S_8	(octasulfur)	
O^{2-}	oxido	oxo†,‡
S^{2-}	sulfido	thio†, thioxo‡
Se^{2-}	selenido	selenoxo‡
Te^{2-}	tellurido	telluroxo‡
$(O_2)^{2-}$	[dioxido(2–)]	peroxo†, peroxy‡
$(O_2)^-$	[dioxido(1–)]	hyperoxo†, superoxido†,‡
$(O_3)^-$	[trioxido(1–)]	ozonido†
$(S_2)^{2-}$	[disulfido(2–)]	(dithio)‡
$(S_5)^{2-}$	[pentasulfido(2–)]	
	(pentasulfane-1,5-diido)	
$(Se_2)^{2-}$	[diselenido(2–)]	(diseleno)‡
$(Te_2)^{2-}$	[ditellurido(2–)]	(ditelluro)‡
H_2O		aqua†,‡
H_2S	(sulfane)	(hydrogen sulfide)†
H_2Se	(selane)	(hydrogen selenide)†
H_2Te	(tellane)	(hydrogen telluride)†
$(OH)^-$	hydroxido	hydroxo†, hydroxy‡
$(SH)^-$	sulfanido, (hydrogensulfido)	
$(SeH)^-$	selanido, (hydrogenselenido)	selenyl‡
$(TeH)^-$	tellanido, (hydrogentellurido)	telluryl‡

Continued on p. 62.

Table 4.6 (*Continued.*)

Formula	Systematic name	Alternative ligand name
H_2O_2		(hydrogen peroxide)
H_2S_2	(disulfane)	(hydrogen disulfide)
H_2Se_2	(diselane)	(hydrogen diselenide)
H_2S_5	(pentasulfane)	(hydrogen pentasulfide)
$(HO_2)^-$		(hydrogenperoxo)†, (hydroperoxy)‡
$(HS_2)^-$	(disulfanido)	(hydrogendisulfido), (hydrodisulfido)‡
$(HS_5)^-$	(pentasulfanido)	(hydrogenpentasulfido)
$(CH_3O)^-$	(methanolato)	methoxo†, methoxy‡
$(C_2H_5O)^-$	(ethanolato)	ethoxo†, ethoxy‡
$(C_3H_7O)^-$	(1-propanolato)	propoxido†, propoxy‡
$(C_4H_9O)^-$	(1-butanolato)	butoxido†, butoxy‡
$(C_5H_{11}O)^-$	(1-pentanolato)	(pentyloxido)†, pentoxy‡
$(C_{12}H_{25}O)^-$	(1-dodecanolato)	(dodecyloxido)†, (dodecyloxy)§
$(CH_3S)^-$	(methanethiolato)	(methylthio)†
$(C_2H_5S)^-$	(ethanethiolato)	
$(C_2H_4ClO)^-$	(2-chloroethanolato)	
$(C_6H_5O)^-$	(phenolato)	phenoxido†, phenoxy‡
$(C_6H_5S)^-$	(benzenethiolato)	(phenylthio)†
$[C_6H_4(NO_2)O]^-$	(4-nitrophenolato)	
CO	(carbon monoxide)	carbonyl†,‡
CS	(carbon monosulfide)	(thiocarbonyl)†, (carbonothioyl)‡
$(C_2O_4)^{2-}$	(ethanedioato)	(oxalato)§
$(HCO_2)^-$	(methanoato)	(formato)§
$(CH_3CO_2)^-$	(ethanoato)	(acetato)§
$(CH_3CH_2CO_2)^-$	(propanoato)	(propionato)§
$(SO_2)^{2-}$	[dioxosulfato(2–)]	[sulfoxylato(2–)]§
$(SO_3)^{2-}$	[trioxosulfato(2–)]	[sulfito(2–)]
$(HSO_3)^-$	[hydrogentrioxosulfato(1–)]	(hydrogensulfito)
$(SeO_2)^{2-}$	[dioxoselenato(2–)]	[selenoxylato(2–)]
$(S_2O_2)^{2-}$	[dioxothiosulfato(2–)]	[thiosulfito(2–)]§
$(S_2O_3)^{2-}$	[trioxothiosulfato(2–)]	[thiosulfato(2–)]
$(SO_4)^{2-}$	[tetraoxosulfato(2–)]	[sulfato(2–)]
$(S_2O_6)^{2-}$	[hexaoxodisulfato(*S—S*)(2–)]	[dithionato(2–)]
$(S_2O_7)^{2-}$	[μ-oxo-hexaoxodisulfato(2–)]	[disulfato(2–)]
$(TeO_6)^{6-}$	[hexaoxotellurato(6–)]	orthotellurato
Ligands based on halogen elements (elements of Group 17)		
Br_2	(dibromine)	
F^-		fluoro
Cl^-		chloro
$(I_3)^-$		[triiodo(1–)]
$[ClF_2]^-$	[difluorochlorato(1–)]	
$[IF_4]^-$	[tetrafluoroiodato(1–)]	
$[IF_6]^-$	[hexafluoroiodato(1–)]	
$(ClO)^-$	[oxochlorato(1–)]	hypochlorito
$(ClO_2)^-$	[dioxochlorato(1–)]	chlorito
$(ClO_3)^-$	[trioxochlorato(1–)	chlorato
$(ClO_4)^-$	[tetraoxochlorato(1–)]	perchlorato
$(IO_5)^{3-}$	[pentaoxoiodato(3–)]	
$(IO_6)^{5-}$	[hexaoxoiodato(5–)]	
$(I_2O_9)^{4-}$	[μ-oxo-octaoxodiiodato(4–)]	

* The use of enclosing marks here is to some extent arbitrary. For example: $[IF_4]^-$ is regarded as a coordination complex, and hence the square brackets; $(ClO_3)^-$ is regarded as a simple anion (Section 4.4.3.1), and hence the parentheses. The braces surrounding systematic names of the ligands are necessary when they are combined into the name of a coordination entity. For organic anions, the sequence of enclosing marks is in accordance with normal organic practice, which differs from the sequence used in coordination nomenclature. Enclosing marks are placed on ligand names as they are to be used in names of coordination entities.
† IUPAC-approved names.
‡ Names used in the *9th Collective Index of Chemical Abstracts*.
§ IUPAC alternative names normally preferred in organic nomenclature.

4.4.3.3   *Designation of donor atom.* In some cases, it may not be evident which atom in a ligand is the donor. This is exemplified by the nitrito ligand in Examples 5 and 10, p. 59. This can conceivably bind through an O or N atom. In simple cases, the donor atom can be indicated by italicised element symbols placed after the specific ligand name and separated from it by a hyphen, as demonstrated in those particular examples. More complex examples will be dealt with below. With polydentate ligands, this device may still be serviceable. Thus, dithiooxalate ion may be attached through S or O, and formulations such as dithiooxalato-*S,S'* and dithiooxalato-*O,O'* should suffice. It could be necessary to use superscripts to the donor atom symbols if these need to be distinguished because there is more than one atom of the same kind to choose from.

Examples

$(CH_3COCHCOCH_3)^-$ pentane-2,4-dionate-C^3

1.
$$
\begin{array}{l}
O{=}C{-}O\diagdown \\
\quad| \qquad\diagup M \\
HCO \\
\quad| \\
HCOH \\
\quad| \\
O{=}C{-}O^-
\end{array}
$$

2.
$$
\begin{array}{l}
O{=}C{-}O^- \\
\quad| \\
HCO \\
\quad| \diagdown M \\
HCO \diagup \\
\quad| \\
O{=}C{-}O^-
\end{array}
$$

3.
$$
\begin{array}{l}
O{=}C{-}O \\
\quad| \qquad\diagdown \\
HCOH \quad\diagdown \\
\quad| \qquad M \\
HCOH \quad\diagup \\
\quad| \qquad\diagup \\
O{=}C{-}O
\end{array}
$$

tartrato(3–)-O^1, O^2 tartrato(4–)-O^2, O^3 tartrato(2–)-O^1, O^4

Complicated examples are more easily dealt with using the kappa convention, and this is particularly useful where a donor atom is part of a group that does not carry a locant according to organic rules. The two oxygen atoms in a carboxylato group demonstrate this. The designator κ is a locant placed after that portion of the ligand name that denotes the particular function in which the ligating atom is found. The ligating atoms are represented by superscript numerals, letters or primes affixed to the donor element symbols, which follow κ without a space. A right superscript to κ denotes the number of identically bound ligating atoms.

Examples

4.

[2-(diphenylphosphino-κ*P*)phenyl-κ*C¹*]hydrido(triphenylphosphine-κ*P*)-nickel(II)

5.

[*N,N'*-bis(2-amino-κ*N*-ethyl)ethane-1,2-diamine-κ*N*]chloroplatinum(II)

6. 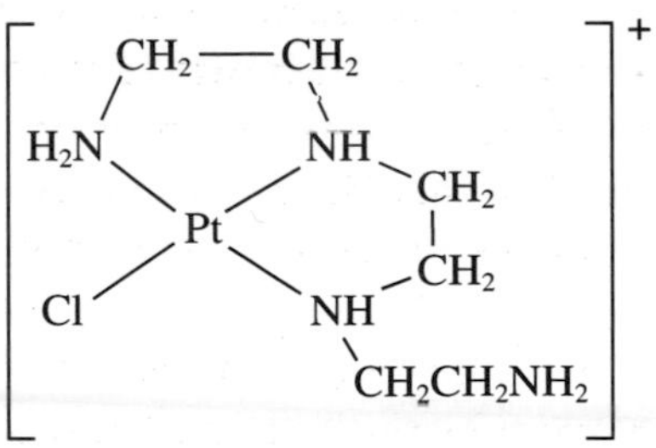 $[N\text{-}(\text{-amino-}\kappa N\text{-ethyl})\text{-}N'\text{-}(2\text{-aminoethyl})\text{-ethane-}$
1,2-diamine-$\kappa^2 N,N'$]chloroplatinum(II)

The two possible didentate modes of binding of 1,4,7-triazacyclononane are shown in the next examples.

Examples

7. 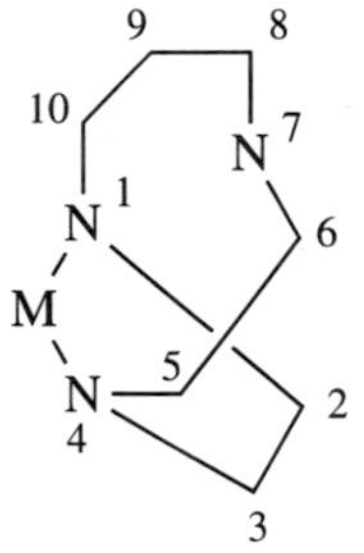 $\kappa^2 N^1, N^4$

8. 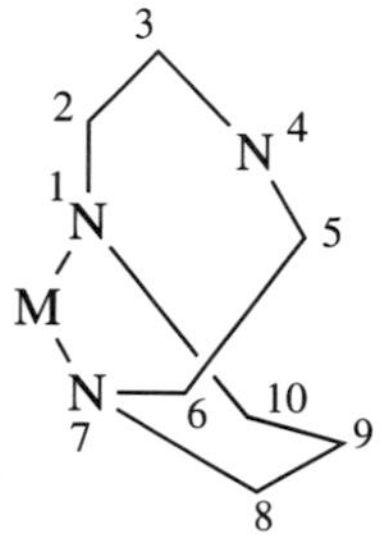 $\kappa^2 N^1, N^7$

The final example is an application to a very complex ligand.

Example

9. 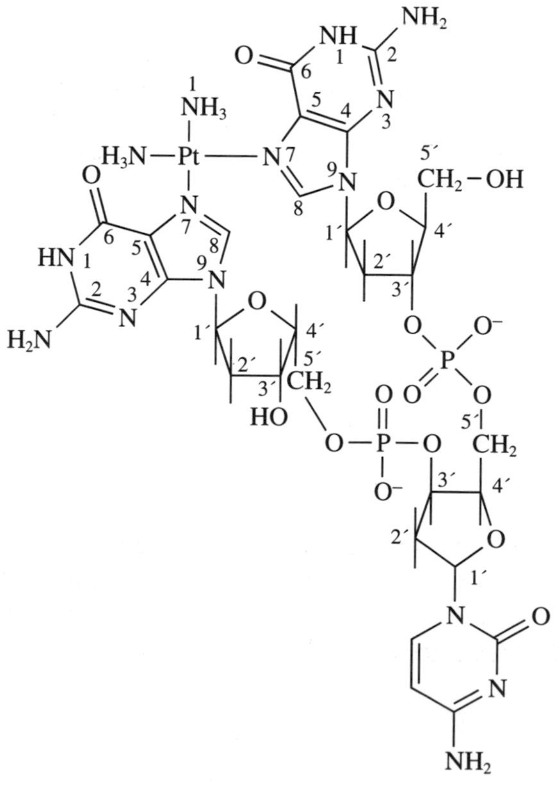 diammine[2′-deoxyguanylyl-
κN^7-(3′→5′)-2′-
deoxycytidylyl(3′→5′)-2′-
deoxyguanosinato(2–)-
κN^7]platinum(II)

4.4.3.4 *Inclusion of structural information.* The names described so far detail ligands and central atoms, but give no information on stereochemistry. The coordination number and shape of the coordination polyhedron may be denoted, if desired, by a polyhedral symbol. These are listed in Table 4.4. Such a symbol is used as an affix in parentheses, and immediately precedes the name, separated from it by a hyphen. This device is not often used.

Geometrical descriptors, such as *cis*, *trans*, *mer* (from meridional) and *fac* (from facial), have found wide usage in coordination nomenclature. The meaning is unequivocal only in simple cases, particularly square planar for the first two and octahedral for the others.

Examples

1. 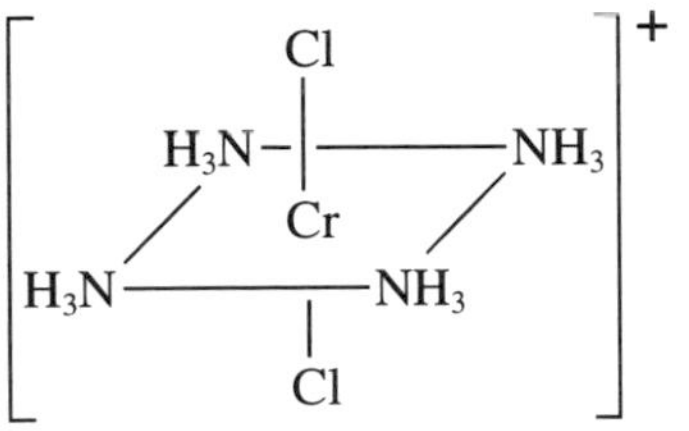 *trans*-isomer
2. 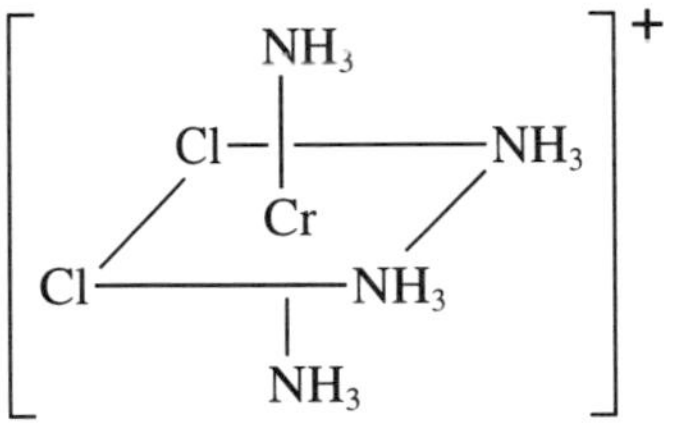 *cis*-isomer
3. 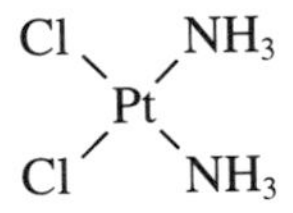 *cis*-isomer
4. 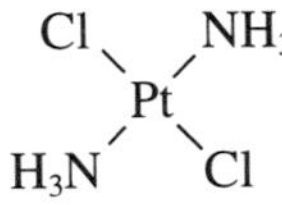 *trans*-isomer
5. 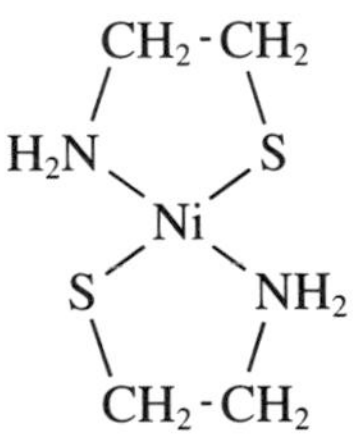 *trans*-isomer
6. 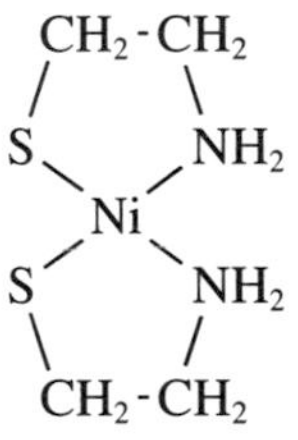 *cis*-isomer
7. 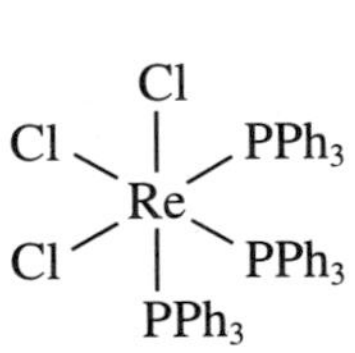 *fac*-isomer
8. 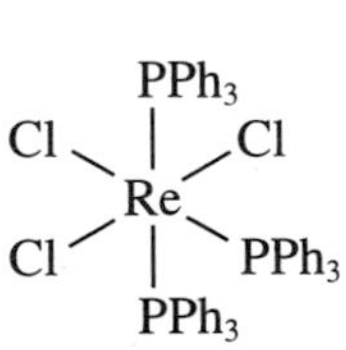 *mer*-isomer

More complex devices have been developed that are capable of dealing with all cases. The reader is referred to the *Nomenclature of Inorganic Chemistry*, Chapter 10. The use of stereochemical descriptors in organic names and formulae is dealt with in Chapter 3, Section 3.8 (p. 21).

4.4.4   **Polynuclear complexes**

4.4.4.1   *Formulae and names.* Polynuclear inorganic complexes may be so complex as to make accurate structure-based formulae and names too complicated to be useful. Compositional formulae and names may be adequate. These are discussed briefly first.

Bridging ligands are indicated by the designator μ. This is followed by the name of the bridging ligand, separated from it by a hyphen. The ligands are cited in alphabetical order, as normal, but the whole term, as in μ-chloro, is separated from the rest of the ligand names by hyphens, e.g. ammine-μ-chloro-chloro. A bridging ligand is cited before a non-bridging ligand of the same kind, as in μ-chloro-chloro. The number of coordination centres connected by the bridge is indicated by the bridge index, a right subscript to μ, as in μ_n. Clearly $n \geqslant 2$, but it is never stated for $n = 2$, and its use for $n > 2$ is optional.

Otherwise, the normal rules expounded for mononuclear complexes apply.

Examples

1. $[Rh_3H_3\{P(OCH_3)_3\}_6]$ trihydridohexakis(trimethyl phosphite)trirhodium

2. $[CoCu_2Sn(CH_3)\{\mu\text{-}(C_2H_3O_2)\}_2(C_5H_5)]$ bis(μ-acetato)(cyclopentadienyl)(methyl)-cobaltdicoppertin

3. $[Fe_2Mo_2S_4(C_6H_5S)_4]^{2-}$ tetrakis(benzenethiolato)tetra-thio(diirondimolybdenum)ate(2–)

If structural information is available, it can be conveyed using the devices already discussed. Bridging ligands are cited as above, unless the symmetry of the system allows a simplification using multiplicative prefixes. Bonding between metal atoms may be indicated in names by italicised element symbols of the appropriate metal atoms, separated by a long dash, and enclosed in parentheses after the list of central atoms and before the ionic charge.

Examples

4. $[\{Cr(NH_3)_5\}_2(\mu\text{-}OH)]Cl_5$ μ-hydroxo-bis(pentaamminechromium)(5+) pentachloride

5. $[PtCl\{P(C_6H_5)_3\}]_2(\mu\text{-}Cl)_2]$ di-μ-chloro-bis[chloro(triphenylphosphine)platinum]

6. $[\{Fe(NO)_2\}_2\{\mu\text{-}P(C_6H_5)_2\}_2]$ bis(μ-diphenylphosphido)bis(dinitrosyliron)

7. $[Br_4ReReBr_4]^{2-}$ bis(tetrabromorhenate)(*Re—Re*)(2–)

8. $[Mn_2(CO)_{10}]$ bis(pentacarbonylmanganese)(*Mn—Mn*), or decacarbonyldimanganese(*Mn—Mn*)

Where the entities to be represented are not symmetrical because, for example, they contain atoms of different metals, an order of citation of metals must be established. In a formula, the priority is established by use of Table IV of the *Nomenclature of Inorganic Chemistry* (Table 3.1 of this book), the highest priority being assigned to the element reached last following the direction of the arrow. In the name, alphabetical order establishes the priority.

Table 4.7 Structural descriptors.

Coordination number of central atom	Descriptor	Point group	Structural descriptor
3	triangulo	D_{3h}	
4	quadro	D_{4h}	
4	tetrahedro	T_d	$[T_d\text{-}(13)\text{-}\Delta^4\text{-}closo]$
5		D_{3h}	$[D_{3h}\text{-}(131)\text{-}\Delta^6\text{-}closo]$
6	octahedro	O_h	$[O_h\text{-}(141)\text{-}\Delta^8\text{-}closo]$
6	triprismo	D_{3h}	
8	antiprismo	S_6	
8	dodecahedro	D_{2d}	$[D_{2d}\text{-}(2222)\text{-}\Delta^6\text{-}closo]$
8	hexahedro (cube)	O_h	
12	icosahedro	I_h	$[I_h\text{-}(1551)\text{-}\Delta^{20}\text{-}closo]$

Examples

9. $[(CO)_5\overset{1}{Re}-\overset{2}{Co}(CO)_4]$
   nonacarbonyl-$1\kappa^5 C$, $2\kappa^4 C$-cobaltrhenium($Co-Re$)

10. $[[\overset{1}{IrCl_2}(CO)\{P(C_6H_5)_3\}_2](\overset{2}{HgCl})]$
    carbonyl-$1\kappa C$-trichloro-$1\kappa^2 Cl$,$2\kappa Cl$-bis(triphenylphosphine-$1\kappa P$)iridiummercury($Hg-Ir$)

11. $[Cr(NH_3)_5(\mu\text{-}OH)Cr(NH_3)_4(NH_2CH_3)]Cl_5$ nonaammine-μ-hydroxo(methanamine)dichromium(5+) pentachloride

12. $[\{Co(NH_3)_3\}_2(\mu\text{-}OH)_2(\mu\text{-}NO_2)]Br_3$ di-μ-hydroxo-μ-nitrito-κN:κO-bis(triamminecobalt)(3+) tribromide

13. $[\{Cu(C_5H_5N)\}_2(\mu\text{-}C_2H_3O_2)_4]$ tetrakis(μ-acetato-κO:$\kappa O'$)bis[(pyridine)copper(II)]

For larger aggregations, a set of structural descriptors (see Table 4.7) is used. Homonuclear entities can have relatively simple names using these descriptors. The examples below give an indication of how names are arrived at. For more complex cases, the reader is referred to the *Nomenclature of Inorganic Chemistry*, p. 192. All the devices already discussed above can be called into use as necessary.

Examples

14. $[\{Co(CO)_3\}_3(\mu_3\text{-}CI)]$
    nonacarbonyl-(μ_3-iodomethylidyne)-*triangulo*-tricobalt(3 $Co-Co$)

15. $Cs_3[Re_3Cl_{12}]$
    caesium dodecachloro-*triangulo*-trirhenate(3 $Re-Re$)(3−)

16. $[Cu_4I_4\{P(C_2H_5)_3\}_4]$
    tetra-μ_3-iodo-tetrakis(triethylphosphine)-*tetrahedro*-tetracopper, or tetra-μ_3-iodo-tetrakis(triethylphosphine)$[T_d\text{-}(13)\text{-}\Delta^4\text{-}closo]$tetracopper

17.

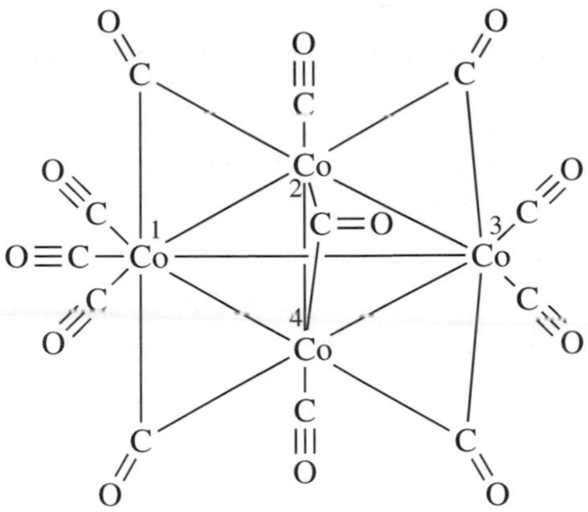

penta-μ-carbonyl-1:2κ^{2}C; 1:4κ^{2}C; 2:3κ^{2}C; 2:4κ^{2}C; 3:4κ^{2}C-heptacarbonyl-1κ^{3}C,2κC,3κ^{2}C,4κC-*tetrahedro*-tetracobalt(4 Co—Co), or penta-μ-carbonyl-1:2κ^{2}C; 1:4κ^{2}C; 2:3κ^{2}C; 2:4κ^{2}C; 3:4κ^{2}C-heptacarbonyl-1κ^{3}C,2κC,3κ^{2}C,4κC-[T_d-(13)-Δ^4-*closo*]tetracobalt(4 Co—Co)

18. [{Hg(CH$_3$)}$_4$(μ$_4$-S)]$^{2+}$

μ$_4$-thio-tetrakis(methylmercury) (2+) ion, or tetramethyl-1κC,2κC,3κC,4κC-μ$_4$-thio-*tetrahedro*-tetramercury(2+) ion, or tetramethyl-1κC,2κC,3κC,4κC-μ$_4$-thio-[T_d-(13)-Δ^4-*closo*]tetramercury(2+) ion

19.

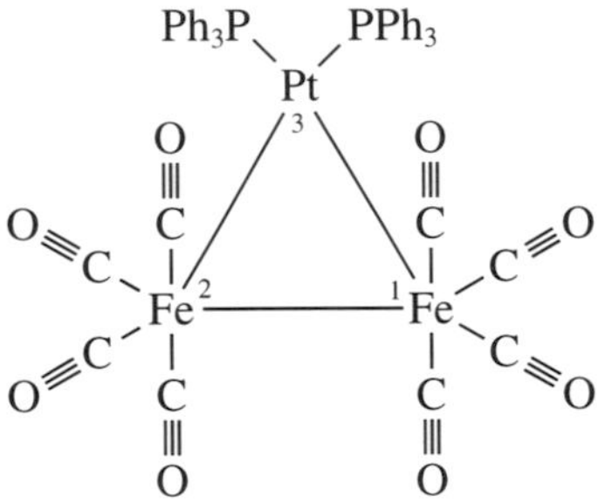

octacarbonyl-1κ^{4}C,2κ^{4}C-bis-(triphenylphosphine-3κP)-*triangulo*-diironplatinum(Fe—Fe) (2 Fe—Pt)

20. [Mo$_6$S$_8$]$^{2-}$

octa-μ$_3$-thio-*octahedro*-hexamolybdate(2–), or octa-μ$_3$-thio-[O_h-(141)-Δ^8-*closo*]hexamolybdate(2–)

21.

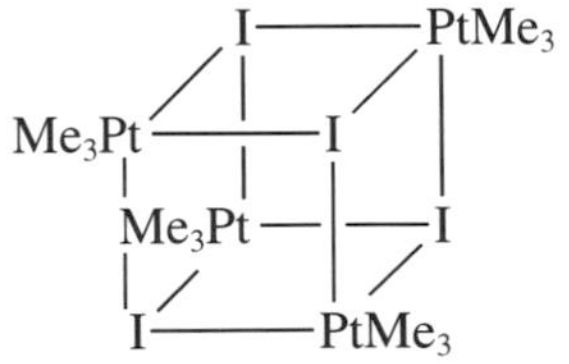

tetra-μ$_3$-iodo-tetrakis[trimethylplatinum(IV)], or tetra-μ$_3$-iodo-dodecamethyl-1κ^{3}C,2κ^{3}C,3κ^{3}C,4κ^{3}C-*tetrahedro*-tetraplatinum(IV), or tetra-μ$_3$-iodo-dodecamethyl-1κ^{3}C,2κ^{3}C,3κ^{3}C,4κ^{3}C-[T_d-(13)-Δ^4-*closo*]tetraplatinum(IV)

22. [Be$_4$(μ-C$_2$H$_3$O$_2$)$_6$(μ$_4$-O)]

hexakis(μ-acetato-κO:κO')-μ$_4$-oxo-*tetrahedro*-tetraberyllium, or hexakis(μ-acetato-κO:κO')-μ$_4$-oxo-[T_d-(13)-Δ^4-*closo*]tetraberyllium

The descriptors [T_d-(13)-Δ^4-*closo*] and [O_h-(141)-Δ^8-*closo*] are useful for precise designations, but simpler names are also available (see the *Nomenclature of Inorganic Chemistry*, p. 192).

4.4.5 **Coordination nomenclature for oxoacids, oxoanions and related compounds**

An oxoacid is a compound that contains oxygen and at least one hydrogen atom attached to oxygen and produces a conjugate base by loss of one or more hydrons. There is a large accumulation of trivial names for such compounds and their derived anions, and these are often incomprehensible without a great deal of rote learning. In addition, the conventional nomenclatures often use suffixes such as -ous and -ic and prefixes such as hypo- and meta- that change in detailed meaning from element to element. For this reason, such nomenclatures are now strongly discouraged.

Coordination nomenclature has been adapted to provide systematic names for many of these species, and several such names have already been quoted in this text. For a more detailed treatment, the reader is referred to the *Nomenclature of Inorganic Chemistry*, Chapter 9. The basic strategy is to treat the acid or anion as a coordination compound, with the central atom or atoms being defined as the coordination centres, whether or not they are metal atoms. The oxygen atoms are then defined as the ligands. The question of whether the 'acid' hydrogen atoms are ionised hydrons or more properly regarded as part of a coordinated hydroxide ion is avoided by citing such hydrogen atoms first in the name, unless there is good empirical reason for not doing so. In the names listed below, the systematic names of the anions are obtained simply by omitting the initial 'acid' hydrogens.

Examples

1.	H_3BO_3	trihydrogen trioxoborate
2.	H_2CO_3	dihydrogen trioxocarbonate
3.	HNO_3	hydrogen trioxonitrate
4.	HNO_2	hydrogen dioxonitrate
5.	HPH_2O_2	hydrogen dihydridodioxophosphate(1–)
6.	H_3PO_3	trihydrogen trioxophosphate(3–)
7.	H_2PHO_3	dihydrogen hydridotrioxophosphate(2–)
8.	H_3PO_4	trihydrogen tetraoxophosphate(V)
9.	$H_4P_2O_7$	tetrahydrogen μ-oxo-hexaoxodiphosphate
10.	$(HPO_3)_n$	poly[hydrogen trioxophosphate(1–)]
11.	$(HO)_2OPPO(OH)_2$	tetrahydrogen hexaoxodiphosphate(*P—P*)(4–)
12.	H_2SO_4	dihydrogen tetraoxosulfate
13.	$H_2S_2O_7$	dihydrogen μ-oxo-hexaoxodisulfate
14.	$H_2S_2O_3$	dihydrogen trioxothiosulfate
15.	$H_2S_2O_6$	dihydrogen hexaoxodisulfate(*S—S*)
16.	$H_2S_2O_4$	dihydrogen tetraoxodisulfate(*S—S*)
17.	H_2SO_3	dihydrogen trioxosulfate

Note the optional use of charge numbers and oxidation states where it is considered helpful.

There is an alternative 'acid' nomenclature that is based upon similar principles, but is not as versatile, and is also used for oxoacids of transition elements, such as tetraoxomanganic acid and μ-oxo-hexaoxochromic acid. It is discussed further in the *Nomenclature of Inorganic Chemistry*, Chapter 9.

The hydrogen nomenclature can also be adapted to yield satisfactory names for derivatives obtained in formal fashion by replacing a coordinated oxo group from the acid. A small selection is given below.

Examples

18.	H_2SO_3S	hydrogen trioxothiosulfate(2–)
19.	HSO_3Cl	hydrogen chlorotrioxosulfate
20.	SO_2Cl_2	dichlorodioxosulfur
21.	$H[PF_6]$	hydrogen hexafluorophosphate(1–)
22.	$SO_2(OCH_3)_2$	dimethoxodioxosulfur
23.	$HOSO_2NH_2$	hydrogen amidotrioxosulfate
24.	$PO(OCH_3)_3$	oxotrimethoxophosphorus
25.	$SO_2(NH_2)_2$	diamidodioxosulfur

These names may not be the only ones used for these compounds, but they are systematic and easily comprehensible.

4.5   SUBSTITUTIVE NOMENCLATURE

4.5.1   Introduction

Substitutive nomenclature was developed using the concepts that governed the development of organic chemistry. However, in nomenclature the term substitution is used in a very restricted sense: only hydrogen atoms can be exchanged with other atoms or groups of atoms. Thus a parent hydride must always be the starting point of a substitution operation. For instance, the two molecules CH_3-Cl and CH_3-OH are always derived from the parent hydride CH_3-H. When atoms other than hydrogen are exchanged, the operation is instead called 'replacement' and the resulting nomenclature called 'replacement nomenclature.'

A substitutive name consists of the name of a parent hydride to which prefixes and suffixes are attached as necessary following the general pattern:

prefixes/name of parent hydride/suffixes

A given organic molecule is generally composed of a carbon skeleton and functional groups. A name matches a structure when the name of the parent hydride corresponds to the skeleton, while prefixes and suffixes represent the functional groups and other structural characteristics, such as geometry.

4.5.2   Alkanes and the basic approach to substitutive names

4.5.2.1   *General.* Alkanes are acyclic hydrocarbons of general formula C_nH_{2n+2}. The carbon atoms are arranged in chains that are either branched or unbranched. Chains are called continuous or unbranched when they are composed of -CH_2- groups with two terminal -CH_3 groups. They are branched when they contain more than two terminal -CH_3 groups. In this case, at least one carbon atom must be joined by single bonds to at least three other carbon atoms.

$$CH_3\text{-}CH_2\text{-}CH_2\text{-}CH_3$$

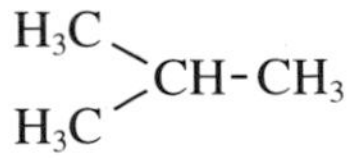

$$\begin{array}{c} H_3C \\ \diagdown \\ \diagup \\ H_3C \end{array} CH\text{-}CH_3$$

an unbranched alkane   a branched alkane

4.5.2.2   *Unbranched alkanes.* Unbranched alkanes are also called normal alkanes. The names of the first four members of the homologous series of unbranched or normal alkanes, C_nH_{2n+2}, are retained names. They were coined more than 100 years ago, officially recognised by the Geneva Conference in 1892, and have been used ever since. There are no alternative names for them.

Examples
1. methane CH_4
2. ethane $CH_3\text{-}CH_3$
3. propane $CH_3\text{-}CH_2\text{-}CH_3$
4. butane $CH_3\text{-}CH_2\text{-}CH_2\text{-}CH_3$

However, higher members of the series are named systematically by combining the ending -ane, characteristic of the first four members and implying complete saturation, with a multiplicative prefix of the series penta-, hexa-, etc. of Table 4.2, which indicates the number of carbon atoms constituting the chain. The letter 'a', which ends the name of the multiplicative prefix, is elided.

Example
5. pent(a) + ane = pentane $CH_3\text{-}[CH_2]_3\text{-}CH_3$

The names of unbranched alkanes are of the utmost importance because these alkanes are the parent hydrides used to name all aliphatic molecules, i.e. molecules having a carbon-chain skeleton.

4.5.2.3   *Branched alkanes.* Branched-chain alkanes can be considered to be constituted of a principal chain and side-chains. They are named by using a precise set of operations:
1 Selection of the principal chain, which will serve as the parent hydride.
2 Identification and naming of side-chains, which will be treated as prefixes.
3 Determination of the position of side-chains on the principal chain and selection of locants using the rule of lowest locants.
4 Selection of the appropriate multiplicative prefixes.
5 Construction of the full name.
     The following example illustrates the step-by-step construction of the name of the branched alkane shown below.

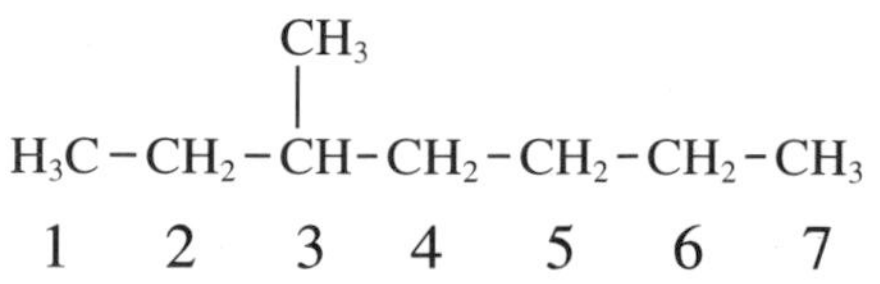

$$\begin{array}{c} CH_3 \\ | \\ H_3C\text{-}CH_2\text{-}CH\text{-}CH_2\text{-}CH_2\text{-}CH_2\text{-}CH_3 \\ 1 \quad\ 2 \quad\ 3 \quad\ 4 \quad\ 5 \quad\ 6 \quad\ 7 \end{array}$$

     The construction of the name begins by selecting the longest chain, which has seven carbon atoms. The disallowed alternatives have four or six. The parent hydride is therefore heptane, and there is one carbon atom in a side-group. This is, of course, a methyl group. Putting these together leads to the partial name methyl-heptane. Multiplicative prefixes are not needed in the present example, as the prefix 'mono' is never used in a substitutive name; finally, the locant '3' is added immediately in front of the part of the name it qualifies: methyl. Locants are separated from other parts of names by hyphens. The full name is 3-methylheptane.

The following three trivial names are still used, but only for the unsubstituted hydrocarbons. Derivatives are named using systematic procedures. These particular names are referred to as retained names.

Examples
1. $(CH_3)_2CH-CH_3$ 2. $(CH_3)_2CH-CH_2-CH_3$ 3. $(CH_3)_4C$

isobutane   isopentane   neopentane

The general characteristics of substitutive nomenclature are now presented in more detail.

4.5.2.4   *Names of alkyl groups.* Unbranched alkyl groups are monovalent groups created by the subtraction of a hydrogen atom from a terminal $-CH_3$ of the unbranched alkane considered to be the parent hydride. They are named by replacing the ending -ane in the name of the parent hydride by -yl. The carbon atom with the free valence always receives the smallest locant, namely '1'. These alkyl groups are called normal or unbranched.

Examples
1. methyl $-CH_3$
2. ethyl $-CH_2-CH_3$
3. propyl $-CH_2-CH_2-CH_3$
4. decyl $-CH_2-[CH_2]_8-CH_3$

Branched alkyl groups are named by prefixing the names of the side-chains to that of the longest unbranched alkyl group.

Example
5. $CH_3-CH_2-CH(CH_3)-CH_2-CH_2-$ 3-methylpentyl
    5   4   3   2   1

The following names are still used, but only for the unsubstituted groups. These particular names are also referred to as retained names. If there are substituents within these groups, systematic procedures must be followed.

Examples
6. $(CH_3)_2CH-$ isopropyl
7. $(CH_3)_2CH-CH_2-$ isobutyl
8. $CH_3-CH_2-CH(CH_3)-$ *sec*-butyl
9. $(CH_3)_3C-$ *tert*-butyl

10. $(CH_3)_2CH\text{-}CH_2\text{-}CH_2-$ isopentyl
11. $CH_3\text{-}CH_2\text{-}C(CH_3)_2-$ *tert*-pentyl
12. $(CH_3)_3C\text{-}CH_2-$ neopentyl

The groups attached to the principal chain are called substituents, and these may be simple or complex. Simple substituents are formed directly from parent hydrides; when a simple substituent is itself substituted, it becomes complex as a consequence. Normal alkyl groups are simple substituents; branched alkyl groups are complex substituents. However, as exceptions the names isopropyl, isobutyl, *sec*-butyl, *tert*-butyl, isopentyl, *tert*-pentyl and neopentyl are taken to refer to simple substituents.

4.5.2.5   *Multiplicative prefixes.* Multiplicative prefixes (Table 4.2) are used when more than one substituent of a given kind is present in a compound or group. The name of the substituent is cited as a prefix, and two sets of multiplicative prefixes are used, depending on whether the substituent is simple or complex.

Basic multiplicative prefixes di-, tri-, tetra-, etc. are used with the names of simple substituents and retained names. Different or modified prefixes are used with complex substituents: bis-, tris-, tetrakis-; from tetrakis- onwards the ending -kis is attached to the basic multiplicative prefix, giving pentakis-, hexakis-, etc. (compare the use in coordination nomenclature).

Examples
1. 3,3-dimethylpentane
2. 5,5-bis(1,2-dimethylpropyl)nonane
3. 4,4-diisopropylheptane or 4,4-bis(1-methylethyl)heptane

In general, chemists like to use retained names. The shorter an approved name, the better.

4.5.2.6   *Lowest locants.* Locants are used to indicate the position of substituents in a compound or group. An almost invariable rule is that locants are selected so that the set used has the lowest possible values. Lowest locants are determined by comparing alternative sets of locants. When compared term-by-term with other locant sets, each in order of increasing magnitude, a set of lowest locants has the lowest term at the first point of difference; for example, 2,3,6,8 is lower than 3,4,6,8 or 2,4,5,7.

Examples

1.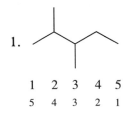

| 1 | 2 | 3 | 4 | 5 |
| 5 | 4 | 3 | 2 | 1 |

2,3-dimethylpentane
(the set 2,3 is *lower* than 3,4)

2.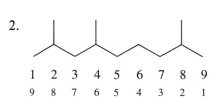

| 1 | 2 | 3 | 4 | 5 | 6 | 7 | 8 | 9 |
| 9 | 8 | 7 | 6 | 5 | 4 | 3 | 2 | 1 |

2,4,8-trimethylnonane
(the set 2,4,8 is *lower* than 2,6,8)

4.5.2.7   *Alphabetical order for citation of detachable prefixes.* Prefixes are used to name substituents, as discussed above. Such prefixes are called detachable prefixes. There is a further class of prefix described as non-detachable. An example is cyclo-, as in cyclohexyl, which is different in stoichiometry from the unmodified hexyl. Non-detachable prefixes are used to modify permanently the name of a parent hydride and thus to create a new parent hydride (see also section 4.5.3.4, p. 78).

    When constructing a name, detachable prefixes are cited in front of the name of the parent hydride in alphabetical order. The names are alphabetised by considering the first letter of each name: 'm' in methyl, 'b' in butyl, 'd' in 1,2-dimethylpropyl. In retained names, the first non-italicised letter is considered: 'i' in isobutyl, 'n' in neopentyl, but 'b' in *tert*-butyl.

    The assembly of the components to construct a full name starts by attaching the names of the detachable prefixes in alphabetical order to the name of the parent hydride. Then, and only then, necessary multiplicative prefixes are introduced, without changing the alphabetical order obtained previously. Finally, locants are inserted.

Examples
1. 4-**e**thyl-2-**m**ethylhexane
2. 4-**e**thyl-2,2-di**m**ethylhexane
3. 6,6-bis(1,2-**d**imethylpropyl)-3,4-di**m**ethylundecane

    In some names with more than one detachable prefix, a set of locants can be attributed in more than one way, as with the locants 3 and 5 in the following example. In such a case, the lowest locant is allocated to the substituent cited first.

Example
4. 3-**e**thyl-5-**m**ethylheptane

As a consequence, the general pattern of substitutive names becomes:

detachable prefixes	non-detachable prefixes	name of parent hydride	suffixes

4.5.2.8   *Criteria for the selection of the principal chain.* It is necessary to lay down rules for the selection of the parent hydride where its identity is not self-evident. The criteria for the selection of the principal chain are listed in Appendix 1. They are very general, as they deal with saturated and unsaturated molecules, and include the use of prefixes and suffixes to characterise all kinds of substituent. In the case of alkanes, neither criterion (a) (there is no principal characteristic group) nor criterion (b) (there is no unsaturation) is relevant. The third criterion (c) is applicable: the principal chain must be the longest. The next relevant criterion (h) is applied when criterion (c) does not permit a definitive choice to be made. According to criterion (h), the principal chain will then be the most substituted amongst all those of equal length that are considered.

Examples

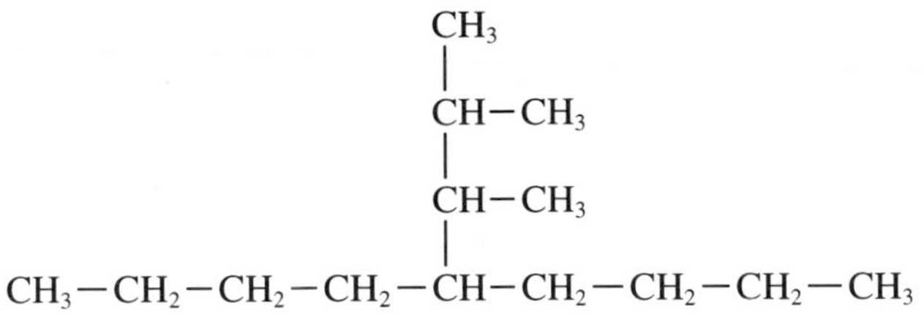

$$CH_3$$
$$|$$
$$CH-CH_3$$
$$|$$
$$CH-CH_3$$
$$|$$
$$CH_3-CH_2-CH_2-CH_2-CH-CH_2-CH_2-CH_2-CH_3$$

5-(1,2-dimethylpropyl)nonane

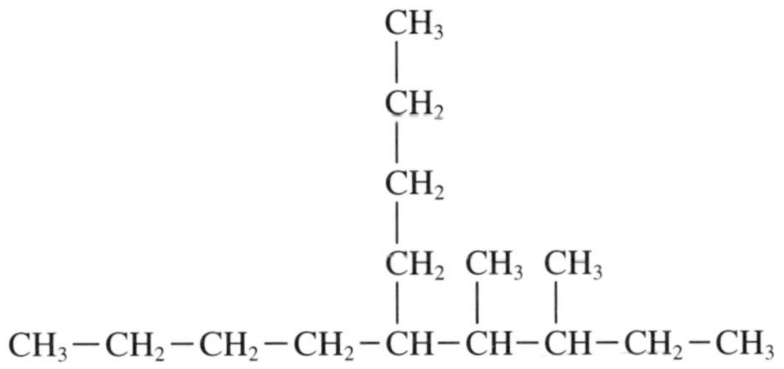

$$CH_3$$
$$|$$
$$CH_2$$
$$|$$
$$CH_2$$
$$|$$
$$CH_2 \quad CH_3 \quad CH_3$$
$$| \quad\quad | \quad\quad |$$
$$CH_3-CH_2-CH_2-CH_2-CH-CH-CH-CH_2-CH_3$$

5-butyl-3,4-dimethylnonane

4.5.3 Cyclic parent hydrides

4.5.3.1 *General.* Continuous-chain alkanes are the sole parent hydrides for all compounds, the skeleton of which is composed of chains. Cyclic parent hydrides are more diverse. In nomenclature, they are classified according to their structure as carbocycles (composed of carbon atoms only) and heterocycles (composed of carbon atoms and other atoms, such as N, O and Si). They are also classified as saturated and unsaturated. Saturated cycles have the maximum number of hydrogen atoms attached to every skeletal atom, as judged by a prespecified valence; unsaturated cycles have fewer hydrogen atoms and multiple bonds between pairs of atoms.

Various degrees and kinds of unsaturation are possible. Unsaturation may be cumulative (which means that there are at least three contiguous carbon atoms joined by double bonds, C=C=C) or non-cumulative (which is another arrangement of two or more double bonds, as in -C=C-C=C-). In nomenclature, unsaturated cyclic parent hydrides have, by convention, the maximum number of non-cumulative double bonds. They are generically referred to as mancudes — derived from the acronym MANCUD, the MAximum Number of non-CUmulative Double bonds. Four classes of cyclic parent hydride are therefore recognised:

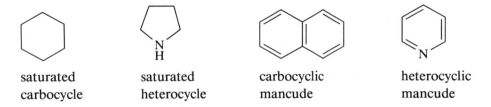

| saturated | saturated | carbocyclic | heterocyclic |
| carbocycle | heterocycle | mancude | mancude |

By the end of 1996 more than 105 000 different rings and ring systems had been encountered (P. M. Giles, Chemical Abstracts Service, personal communication).

For comparison, only a few dozen alkanes are designated as parent hydrides. Over 73 000 rings and ring systems are mancudes and about 10 000 are carbocycles.

A second general feature of this nomenclature is the use of prefixes to modify the names of basic parent hydrides in order to create new parent hydrides. Consequently, these prefixes are non-detachable, permanently attached to the name of the basic parent hydride, and are treated in alphabetic procedures like non-detachable prefixes in alkane nomenclature. Among other things, these non-detachable prefixes are used to indicate the conversion of a chain to a ring (e.g. cyclo-, as in cyclohexane; see Section 4.5.3.2), the opening of rings (e.g. seco-, as in some natural product names; see Chapter 7, section 7.5, p. 122), the fusion of rings (e.g. benzo-, as in benzopyrene) and the replacement of carbon atoms in rings by heteroatoms, thus transforming carbocycles into heterocycles (e.g. phospha-, as in phosphacycloundecane; see Section 4.5.3.3, p. 77).

4.5.3.2   *Monocyclic parent hydrides.* Saturated monocyclic carbocycles are generically called cycloalkanes. Individual rings are named by adding the non-detachable prefix cyclo- to the name of the normal alkane having the same number of carbon atoms. The carbocyclic mancude having six carbon atoms is named benzene. Higher mancude homologues having the general formula C_nH_n or C_nH_{n+1} are named [x]annulenes, x representing the number of carbon atoms in the ring. Annulenes with an odd number of carbon atoms are further characterised by the symbol H to signal the presence of a special hydrogen atom called 'indicated hydrogen'. This symbol is a non-detachable prefix.

Examples

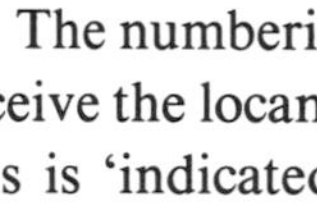

| cyclohexane | benzene | 1H-[7]annulene | [8]annulene |

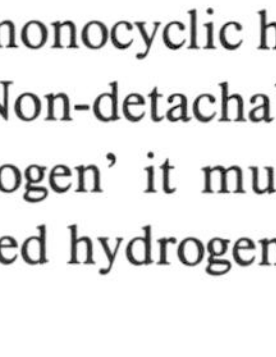

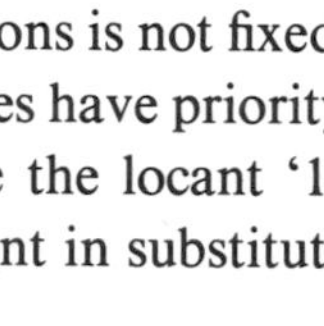

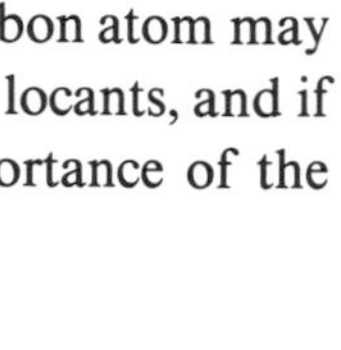

The numbering of monocyclic hydrocarbons is not fixed, as any carbon atom may receive the locant '1'. Non-detachable prefixes have priority for lowest locants, and if this is 'indicated hydrogen' it must receive the locant '1'. The importance of the presence of an 'indicated hydrogen' is evident in substituted rings.

Examples

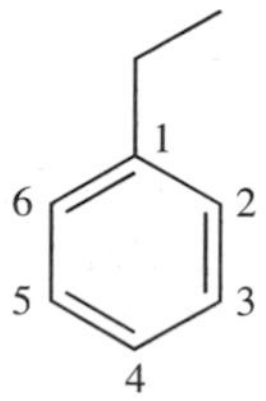

ethylbenzene   2-methyl-1H-[7]annulene

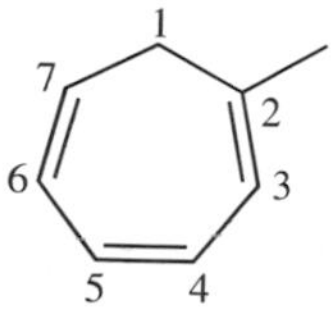

4.5.3.3   *Heterocyclic parent hydrides.* These compounds form a large and diverse group. The names of these parent hydrides are usually formed systematically. However, some 50 trivial names are retained and used in preference to their systematic counterparts.

Examples

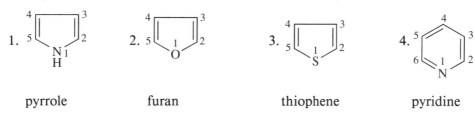

| pyrrole | furan | thiophene | pyridine |

The two most important methods for naming heterocyclic parent hydrides are the extended Hantzsch–Widman system and replacement nomenclature.

The extended Hantzsch–Widman system is used to name monocycles of saturated and mancude rings with between three and ten members, inclusive. The names are composed of two parts: non-detachable prefix(es), indicating the heteroatom; and a stem indicating the size of the ring. Names of prefixes (called 'a' prefixes) are listed in Table 4.8 and names of stems in Table 4.9.

The choice of a stem corresponding to the groups of compounds with six ring atoms and designated 6A, 6B or 6C in Table 4.9 is determined by the atom, the name of which immediately precedes the name of the stem:

6A O,S,Se,Te,Bi,Hg   6B N,Si,Ge,Sn,Pb   6C B,P,As,Sb

Names are formed by eliding the final letter 'a' of the 'a' prefix before it is attached to the stem.

When one heteroatom is present, the locant '1' is attributed to the heteroatom. In many cases, an 'indicated hydrogen' is necessary to describe the structure with accuracy. The presence of more than one heteroatom of any type is indicated by a multiplicative prefix (di-, tri-, etc.). If two or more kinds of heteroatom occur in the same name, the order of citation is the order of their appearance in descending Table 4.8. The numbering starts from the heteroatom cited first in Table 4.8, giving the lowest possible locants to other heteroatoms.

Examples

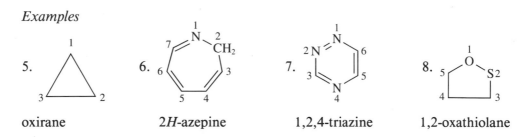

oxirane   2H-azepine   1,2,4-triazine   1,2-oxathiolane

Replacement nomenclature is used to name heteromonocycles that contain more than ten atoms. In developing a replacement name, carbon atoms are regarded as exchanged for heteroatoms. The non-detachable prefixes (Table 4.8) are used to indicate the exchange. Cycloalkane or annulene names are the bases for transformation into the name of a heterocycle.

Table 4.8 The 'a' prefixes used in Hantzsch–Widman nomenclature.

Element	Oxidation state	Prefix	Element	Oxidation state	Prefix
Oxygen	II	oxa	Bismuth	III	bisma
Sulfur	II	thia	Silicon	IV	sila
Selenium	II	selena	Germanium	IV	germa
Tellurium	II	tellura	Tin	IV	stanna
Nitrogen	III	aza	Lead	IV	plumba
Phosphorus	III	phospha	Boron	III	bora
Arsenic	III	arsa	Mercury	II	mercura
Antimony	III	stiba			

Examples

9. phosphacycloundecane
10. sila[12]annulene

4.5.3.4   *Polycyclic parent hydrides.* These are classified as bridged polyalkanes (also known as von Baeyer bridged systems, from the nomenclature system developed to name them), spiro compounds, fused polycyclic systems and assemblies of identical rings. The four systems may be either carbocyclic or heterocyclic. In developing their names, the following principles are used.

• The non-detachable prefixes bicyclo-, tricyclo-, etc. and spiro- characterise the bridged and the spiro systems. Numbers in square brackets give necessary information about the lengths and positions of the bridges in these polycyclic systems.

Table 4.9 Name stems used in Hantzsch–Widman nomenclature.

Ring size	Unsaturated (mancude)	Saturated (mancude)
3	-irene*	-irane†
4	-ete	-etane†
5	-ole	-olane†
6A‡	-ine	-ane
6B‡	-ine	-inane
6C‡	-inine	-inane
7	-epine	-epane
8	-ocine	-ocane
9	-onine	-onane
10	-ecine	-ecane

* The traditional stem 'irine' may be used for rings containing nitrogen only.

† The traditional stems 'iridine', 'etidine' and 'olidine' are preferred for rings containing nitrogen.

‡ For ring size six, the terminations 6A refer to compounds of O, S, Se, Te, Bi and Hg; the terminations 6B refer to compounds of N, Si, Ge, Sn and Pb; and the terminations 6C refer to compounds of B, P, As and Sb.

Examples

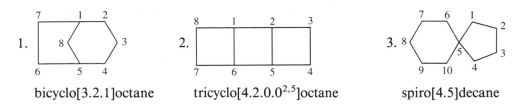

1. bicyclo[3.2.1]octane   2. tricyclo[4.2.0.0^{2,5}]octane   3. spiro[4.5]decane

Heterocyclic systems, which can be regarded as formed by replacement of carbon atoms in the parents described above by heteroatoms, are named by replacement nomenclature.

Examples

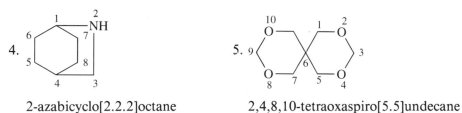

4. 2-azabicyclo[2.2.2]octane   5. 2,4,8,10-tetraoxaspiro[5.5]undecane

• Fused polycyclic systems are very numerous and diverse. They are named using the structures and names of their smaller components and the concept of *ortho*-fusion, which is purely a formal operation encountered in nomenclature. This concept is essential to the naming of larger systems, and is the formation of one bond by the condensation of two bonds belonging to two different cyclic systems, one of them being a mancude ring.

About 60 fused polycyclic systems have trivial names that are retained for present use.

Examples

6. 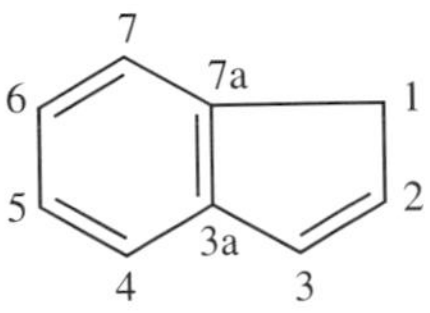 indene   7. 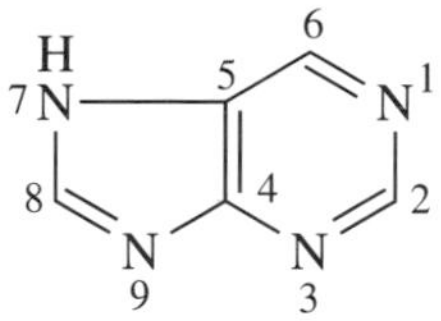 purine
(special numbering system)   8. 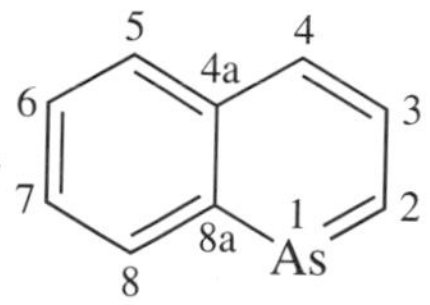 arsinoline   9. 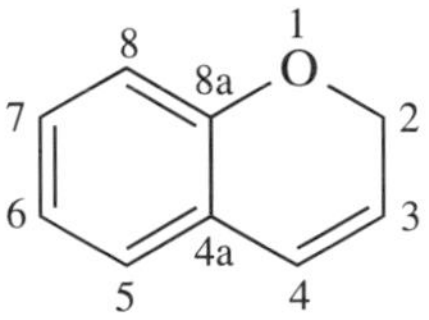 2*H*-chromene

Some fused systems are named systematically by using a multiplicative prefix in front of an ending representing a well-defined arrangement of cycles. For instance, the ending -acene, taken from anthracene, indicates a linear arrangement of benzene rings, as in tetracene, pentacene, etc.

Examples

10.

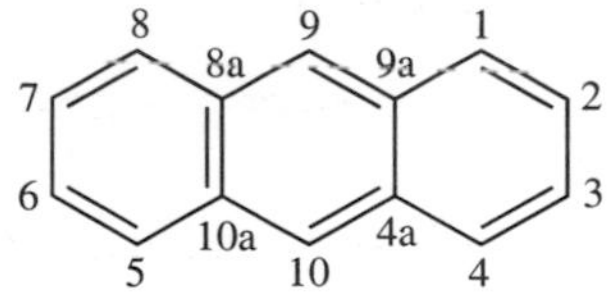

anthracene

11.

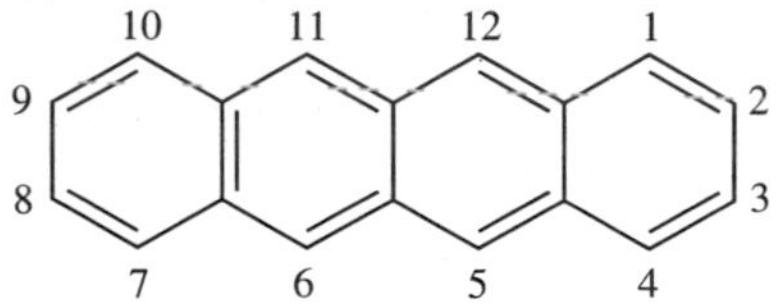

tetracene

12.

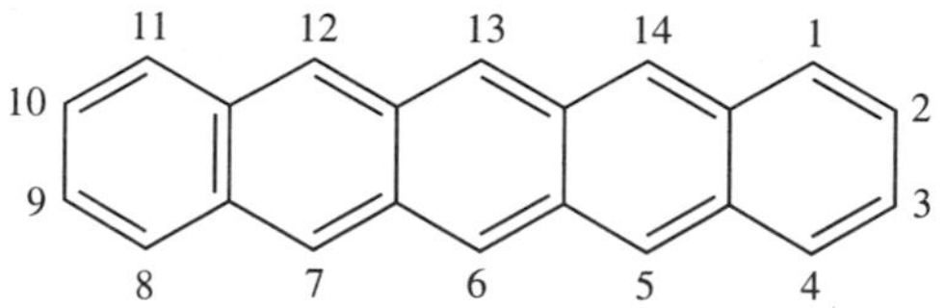

pentacene

When a system does not have a retained name or a name that can be composed systematically as above, and when *ortho*- and *ortho*–*peri*-fusion are possible, it is named using fusion nomenclature, i.e. by combining the names of the two or more systems that are fused. One system is adjudged to be the senior according to criteria described elsewhere and is taken as a parent hydride, and the other is denoted in the name by a non-detachable prefix. The junction of the two systems is indicated in a specific manner. Instead of numerical locants, italic letters '*a*', '*b*', '*c*', etc. are used to identify bonds in the parent hydride. The final letter 'o' and normal locants are characteristic of the prefix. The examples below illustrate the fusion operation and the resulting fusion name.

Examples

13.

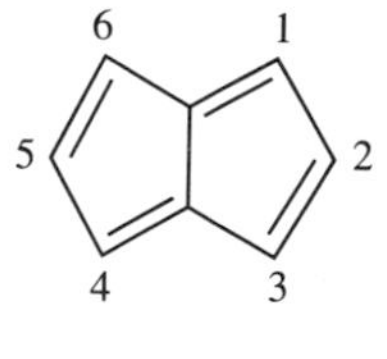

pentalene

14.

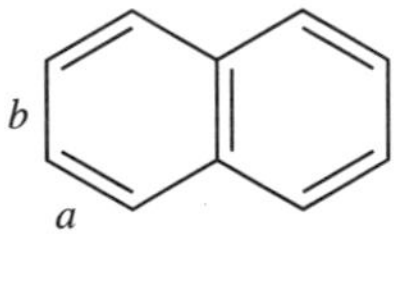

naphthalene

15.

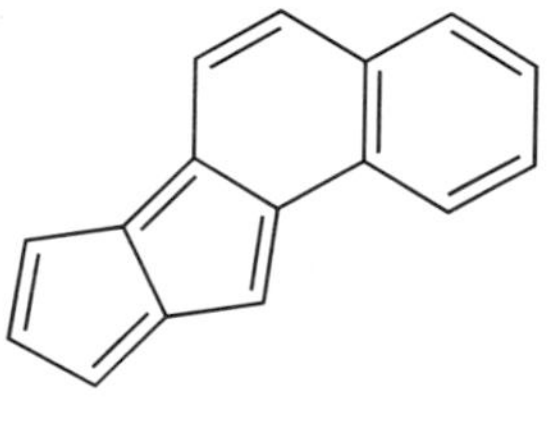

pentaleno[2,1-*a*]naphthalene

• Assemblies of identical rings are named by using a unique set of multiplicative prefixes, bi-, ter-, quater-, etc., to indicate the number of rings. The name phenyl is used instead of benzene for the aromatic C_6 ring.

Examples

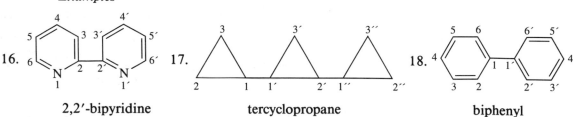

16.

2,2′-bipyridine

17.

tercyclopropane

18.

biphenyl

4.5.4   **Names for substances with various degrees of saturation**

4.5.4.1   *General.* Two methods are used to indicate the degree of saturation in a compound, depending on the nature of the parent hydride.

In one method, the non-detachable prefixes hydro- and dehydro- (in the case of mancudes only) can be used to indicate the addition or subtraction of one hydrogen atom. As even numbers of hydrogen atoms are involved when a carbon–carbon single bond becomes a double or triple bond, the multiplicative prefixes di-, tetra-, etc., as well as the appropriate locants, are used. The prefix dehydro- is always used to indicate the subtraction of hydrogen atoms from saturated heterocycles having Hantzsch–Widman names.

Examples

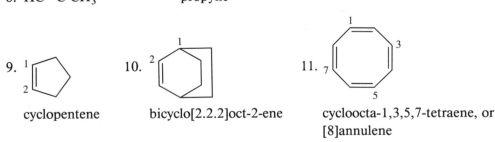

1.   2.   3.

1,2-dihydroazulene   1,4-dihydrophosphinine   1,2-didehydrobenzene
                                                                 (formerly benzyne)

In the other method, the ending -ane is changed to -ene or -yne to indicate the presence of a double or triple bond. This is used for alkanes and mono- and polycyclic alkane parent hydrides. In alkanes and cycloalkanes, the change of the -ane ending to -ene or -yne indicates the presence of one double or triple bond. Multiplicative prefixes di-, tri-, tetra-, etc. are used to signal the multiplicity of unsaturated bonds. Locants placed immediately in front of the endings -ene and -yne are used as needed.

Examples

4. $CH_2{=}CH_2$ ethene (ethylene is no longer approved)
5. $CH_2{=}CH{-}CH_2{-}CH_3$ but-1-ene
6. $CH_2{=}CH{-}CH{=}CH_2$ buta-1,3-diene
7. $HC{\equiv}CH$ acetylene (retained trivial name in addition to the systematic ethyne)
8. $HC{\equiv}C{-}CH_3$ propyne

9.   10.   11.

     cyclopentene   bicyclo[2.2.2]oct-2-ene   cycloocta-1,3,5,7-tetraene, or
                                                                     [8]annulene

If double and triple bonds are present in a structure, they are considered together when assigning lowest locants. Only when this does not allow a resolution do double bonds receive the lowest locants. In a name, the ending -ene is cited before -yne, but with elision of the final 'e'.

Examples

12. $HC\equiv C\text{-}CH=CH\text{-}CH_3$ pent-3-en-1-yne
13. $HC\equiv C\text{-}CH=CH_2$ but-1-en-3-yne

4.5.4.2   *Unsaturated and divalent substituents.* Unsaturated monovalent substituents are named systematically by attaching the suffixes -yl to the stem of the parent name that carries the ending -ene or -yne. Retained trivial names include vinyl for $-CH=CH_2$, allyl for $-CH_2\text{-}CH=CH_2$ and isopropenyl for $-C(CH_3)=CH_2$, but only when it is unsubstituted.

Examples

1. $-CH_2\text{-}CH_2\text{-}CH=CH\text{-}CH_3$ pent-3-en-1-yl
2. $-CH_2\text{-}C\equiv CH$ prop-2-yn-1-yl

Note that the position of the free valence is always given the locant 1.

Divalent substituents of the type $R\text{-}CH=$ or $R_2C=$ are named by appending the suffix -ylidene to an appropriate stem. When the two free valencies are on different atoms or of the type R-CH< or $R_2=C<$ and not involved in the same double bond, the composite suffix -diyl (di- + -yl) is used. Retained names include methylene for $-CH_2\text{-}$, ethylene for $-CH_2\text{-}CH_2\text{-}$ and isopropylidene for $=C(CH_3)_2$, but only when it is unsubstituted.

Examples

3. methylidene $=CH_2$
4. ethylidene $=CH\text{-}CH_3$
5. ethane-1,1-diyl $>CH\text{-}CH_3$
6. propane-1,3-diyl $-CH_2\text{-}CH_2\text{-}CH_2-$

4.5.4.3   *Selection of the principal chain in unsaturated branched acyclic hydrocarbons.* A principal chain must be chosen upon which to base the name of branched unsaturated acyclic hydrocarbons. The general criteria listed in Table 4.10 are applied. Those that are specifically relevant to polyenes and polyynes are: criterion (b), which is the maximum number of double and triple bonds considered together; criterion (c), which is the maximum length; criteria (d), (f) and (g); criterion (h), which is the maximum number of substituents cited as prefixes; and criteria (i) and (j).

Examples

1.
$$H_2C=CH-\underset{\underset{C_6H_{13}}{|}}{C}=CH_2$$

2-hexylbuta-1,3-diene
(criterion b)

2.
$$H_2C=\underset{\underset{C_2H_5}{|}}{C}-CH_2\text{-}CH_2\text{-}CH_2\text{-}CH_3$$

2-ethylhex-1-ene
(criteria b and c)

3.

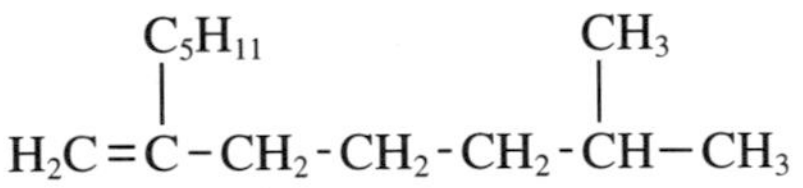

$$C_5H_{11} \qquad\qquad CH_3$$
$$H_2C\!=\!C\!-\!CH_2\!-\!CH_2\!-\!CH_2\!-\!CH\!-\!CH_3$$

6-methyl-2-pentylhept-1-ene
(criteria b, c and h)

Table 4.10 Seniority of chains (the principal chain)

When a choice has to be made of principal chain in an acyclic compound, the following criteria are applied successively, in the order listed, until a single chain is left under consideration. This is then the principal chain.

(a) Select the chain(s) that has (have) the maximum number of substituents corresponding to the principal group.
(b) If this is not definitive, select the chain that has the maximum number of double and triple bonds, considered together.

If (a) and (b) together are not definitive, the following criteria are then applied in order until only one chain remains under consideration.

(c) Chain with maximum length.
(d) Chain with maximum number of double bonds.
(e) Chain with lowest locants for the principal groups, which are those cited as suffixes.
(f) Chain with lowest locants for multiple bonds.
(g) Chain with lowest locants for double bonds.
(h) Chain with maximum number of substituents cited as prefixes.
(i) Chain with lowest locants for all substituents in the principal chain cited as prefixes.
(j) Chain with the substituent that comes first in alphabetical order.

4.5.5 Assemblies of different rings, and of rings with chains

Assemblies of different rings are given substitutive names in which one ring is chosen as the parent hydride, and the other is denoted by a prefix.

The names of mono- or divalent cyclic substituents are formed by adding the suffix -yl or -ylidene, as appropriate, to the name of the cycle, with the exception of cycloalkanes, for which the suffixes replace the ending -ane rather than adding to it. Note that designations such as >—{ and C—{ are used to indicate a free valence whenever >— and C— might otherwise be taken to indicate an appended methyl group.

Examples

1. 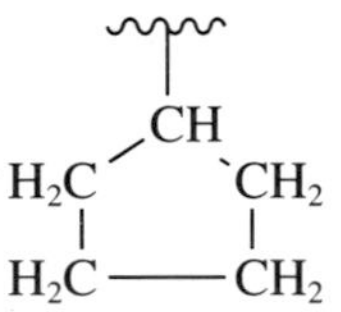 cyclopentyl   2. 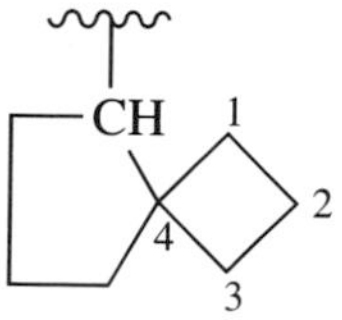 spiro[3.4]octan-5-yl   3. 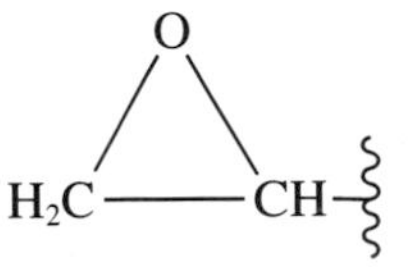 oxiran-2-yl   4. 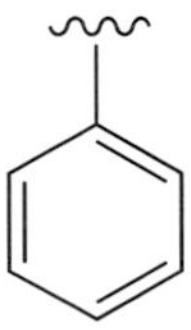 phenyl

5. 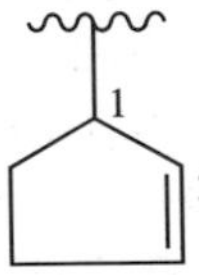 cyclopent-2-en-1-yl

6. 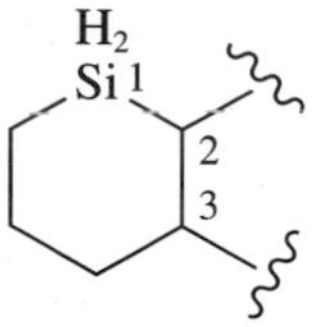 silinane-2,3-diyl

7. 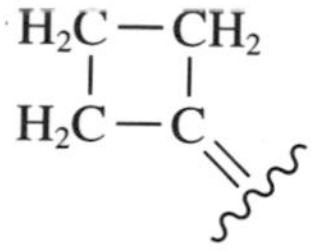 cyclobutylidene

The principal ring is selected using the criteria in Table 6.1. For instance, heterocycles have priority over carbocycles and unsaturated systems have priority over saturated systems.

Examples

8. 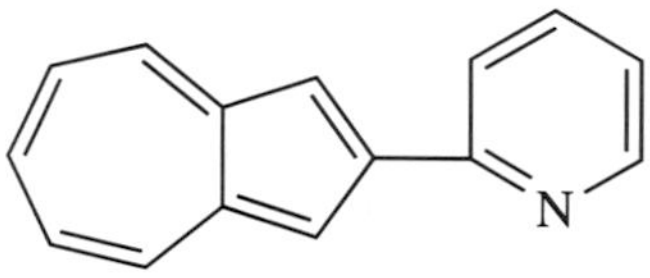 2-(azulen-2-yl)pyridine

9. 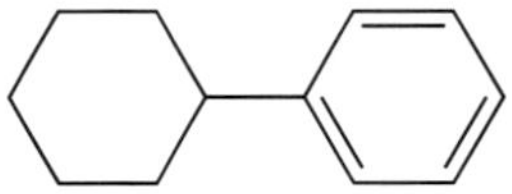 cyclohexylbenzene

Compounds composed of rings and chains are also named using substitutive nomenclature. The ring, whether carbocyclic or heterocyclic, is always selected to be the parent hydride.

Examples

10. $(C_6H_5)_2CH_2$

1,1'-methylenedibenzene

11. 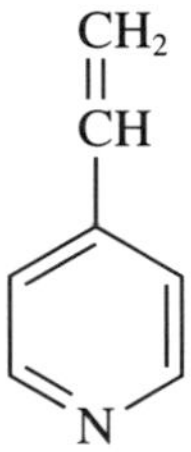 4-vinylpyridine

12. 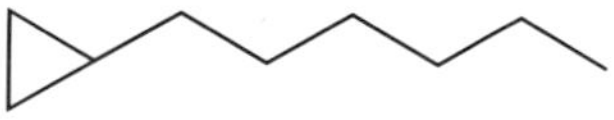 hexylcyclopropane

Trivial names of substituted benzenes retained for present use include toluene, styrene and stilbene, but only for derivatives when substituting in the ring (see Section 4.5.6.2 and Table 4.13).

4.5.6   **Nomenclature of functionalised parent hydrides**

4.5.6.1   *The use of suffixes and prefixes.* The groups that are typical of the various classes of organic compound, such as

-OH in alcohols, >C=O in ketones and $-C\begin{smallmatrix}O\\OH\end{smallmatrix}$ in carboxylic acids

are called characteristic groups or functions.

The names of characteristic groups are cited as suffixes or prefixes when these classes of organic compound are named using substitutive nomenclature. If only one characteristic group is present, then its name is cited as a suffix. If more than one type of characteristic group is present, one must be chosen as the principal group and its name is then cited as a suffix. The names of all other characteristic groups are cited as detachable prefixes. However, some characteristic groups are expressed in a name as prefixes only and are never cited as suffixes. Suffixes and prefixes are listed in Tables 4.11 and 4.12. The detailed usage is exemplified in the following discussion.

Table 4.11 Some characteristic groups with names always cited as prefixes in substitutive nomenclature.

Class	Group	Prefix	Class	Group	Prefix
Halogen compounds	F	fluoro-	Ethers	O-R	R-oxy-
	Cl	chloro-	Nitro compounds	NO_2	nitro-
	Br	bromo-			
	I	iodo-	Nitroso compounds	NO	nitroso-

Table 4.12 Some characteristic groups* with names cited as suffixes or prefixes in substitutive nomenclature, presented in decreasing order of seniority (see Tables 4.10 and 6.1).

Class	Suffix	Prefix	Group
Radicals Anions Cations	see Table 4.14		
Carboxylic acids	-oic acid	carboxy-†	-(C)OOH
	-carboxylic acid	carboxy-	-COOH
Sulfonic acids	-sulfonic acid	sulfo-	$-SO_2OH$
Esters	—‡	(R-oxy)-oxo-	-(C)OOR
		R-oxycarbonyl-	-COOR
Acyl halides	—‡	halo-oxo-	-(C)OHal
		halocarbonyl-	-COHal
Amides	-amide	amino-oxo-	$-(C)ONH_2$
	-carboxamide	aminocarbonyl-	$-CONH_2$
Nitriles	-nitrile	cyano-†	-(C)N
	-carbonitrile	cyano-	-CN
Aldehydes	-al	oxo-	-(C)HO
	-carbaldehyde	formyl-	-CHO
Ketones	-one	oxo-	=O
Alcohols, phenols	-ol	hydroxy-	-OH

* The carbon atoms in parentheses in some groups belong to the parent hydride, normally a chain. If there are no parentheses in the formula, the name of the group as a whole is added to the name of the parent hydride.
† When attached to a ring, the groups -COOH and -CN are never treated as -(C)OOH or -(C)N. If an unbranched chain is directly linked to two or more carboxy groups, the name is based upon that chain and the carboxy groups are treated as substituents -COOH and not as (C)OOH.
‡ See discussion of functional class nomenclature, Chapter 4, Section 4.6 (p. 96).

4.5.6.2   *Names of characteristic groups always cited as prefixes.* The names of the two polyhalogenated ethanes, a and b, are a good illustration of the application of the rule of lowest locants and of the alphabetical order to assign lowest locants.

Examples
1. $ClF_2C-CHBrI$ a   2-bromo-1-chloro-1,1-difluoro-2-iodoethane
2. $BrF_2C-CClFI$ b   1-bromo-2-chloro-1,1,2-trifluoro-2-iodoethane

In monocyclic hydrocarbons, the locant '1' is omitted, but it is necessary in polysubstituted compounds.

Examples

3. 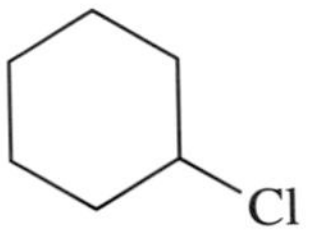   chlorocyclohexane

4. 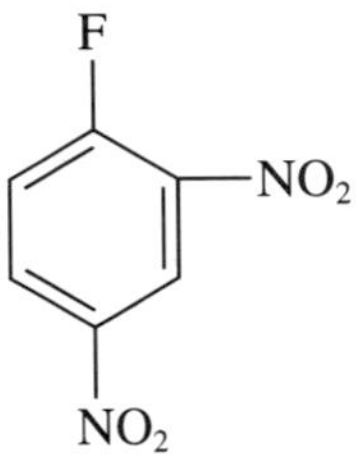   1-fluoro-2,4-dinitrobenzene

5. 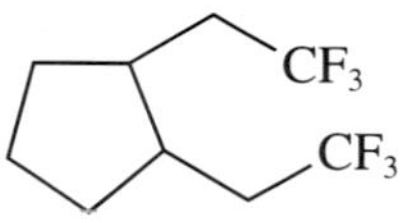   1,2-bis(2,2,2-trifluoroethyl)cyclopentane

When the numbering is predetermined by the nature of the parent hydride, as in polycyclic hydrocarbons and in heterocyclic compounds, lowest locants are still the rule.

Examples

6. 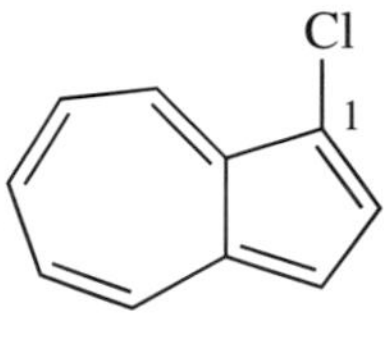   1-chloroazulene

7. 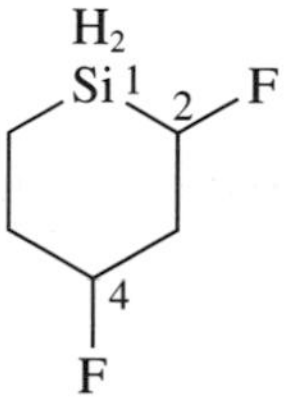   2,4-difluorosilinane

8.

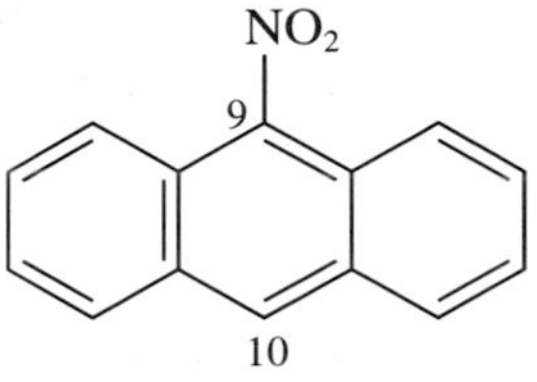

9-nitroanthracene

If a choice has to be made, lowest locants are assigned first to heteroatoms in cycles, then to positions of unsaturation and, finally, to substituents cited as detachable prefixes.

Examples

9.

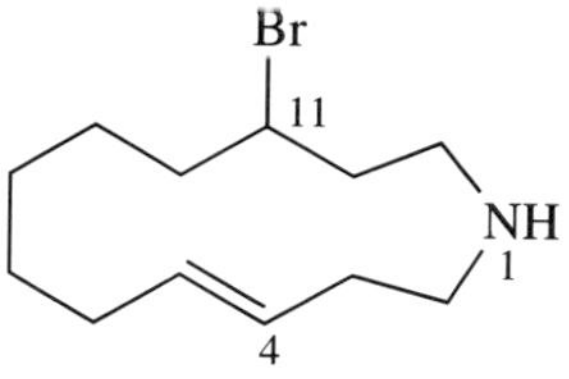

11-bromo-1-azacyclotridec-4-ene

10.

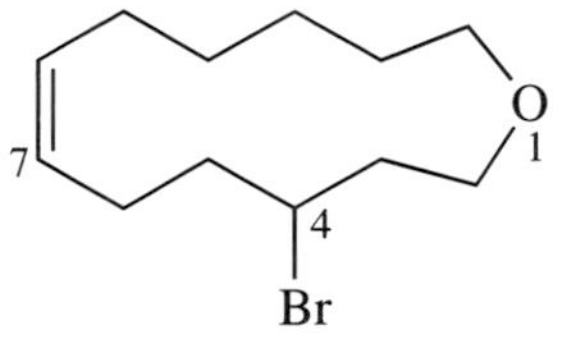

4-bromo-1-oxacyclotridec-7-ene

Normally, the names of all substituents are cited as prefixes in front of the name of the parent hydride, but there are three exceptions. The names of three substituted benzenes — toluene, styrene and stilbene — are retained and can be used to name substituted derivatives, as long as the substitution is only on the ring.

Examples

11.

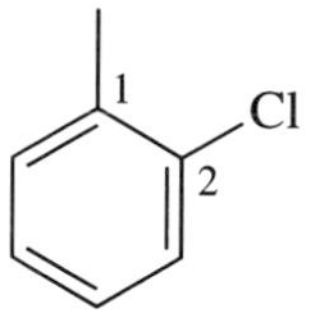

2-chlorotoluene

12.

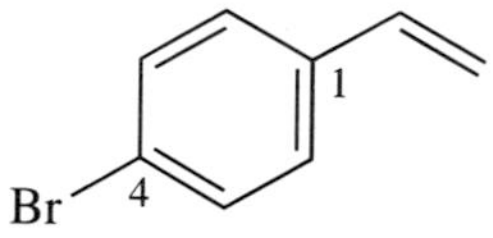

4-bromostyrene

13.

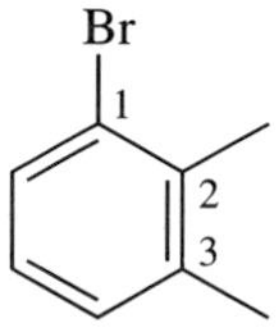

1-bromo-2,3-dimethylbenzene (not 3-bromo-*o*-xylene)

4.5.6.3   *Names of characteristic groups cited as suffixes or prefixes.* In substitutive nomenclature, a suffix must be used whenever possible for the preferred functional group. Prefixes are used to name all characteristic groups except principal functional groups. Lowest locants and multiplicative prefixes di-, tri-, tetra-, etc. are used following the general rules stated in Section 4.5.2 (p. 70). Suffixes and prefixes are listed in Tables 4.11 and 4.12. Monofunctional compounds are named as follows:

1. $CH_3\text{-}CH_2\text{-}CH_2OH$ propan-1-ol
2. $CH_3\text{-}CO\text{-}CH_2\text{-}CH_2\text{-}CH_3$ pentan-2-one
3. $CH_3\text{-}CH(OH)\text{-}CH(OH)\text{-}CH_3$ butane-2,3-diol

4.

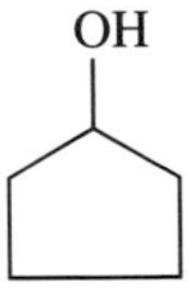

OH

cyclopentanol

The suffixes -oic acid, -al, -amide and -nitrile are used to name acyclic compounds having one or two characteristic groups. Locants are not necessary, as these groups must be at the end of a chain. The suffixes -carboxylic acid, -carbaldehyde, -carboxamide and -carbonitrile are used when more than two groups are attached to chains or one or more groups are attached to cycles.

Examples
5. $CH_3\text{-}CH_2\text{-}COOH$ propanoic acid
6. $OHC\text{-}CH_2\text{-}CH_2\text{-}CHO$ butanedial
7. $CH_2(COOH)\text{-}CH(COOH)\text{-}CH_2(COOH)$ propane-1,2,3-tricarboxylic acid

8.

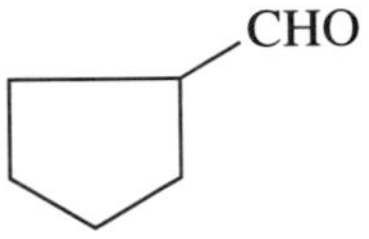

CHO

cyclopentanecarbaldehyde

9. $CH_3\text{-}[CH_2]_3\text{-}CONH_2$ pentanamide
10. $NC\text{-}[CH_2]_4\text{-}CN$ hexanedinitrile

11.

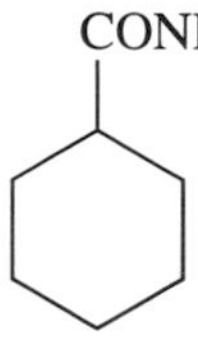

CONH₂

cyclohexanecarboxamide

12.

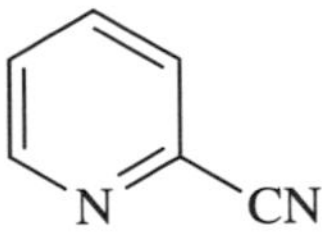

CN N

pyridine-2-carbonitrile

Suffixes and prefixes are necessary to name structures with discontinuities, for instance when characteristic groups are situated on side-chains, or when the carbon skeleton is composed of rings and chains. A principal component, ring or chain, must be chosen. The principal chain is chosen in accordance with the selection criteria listed in Table 4.10 and is the chain supporting the greatest number of characteristic groups. If this is not decisive, the longest chain — criteria (a) and (c) — is chosen.

Example

13.

$$CH_2OH$$
$$HOCH_2-CH_2-CH-CH_2-CH_2-CH_2-CH_2-CH_2OH$$

3-(hydroxymethyl)octane-1,8-diol

In the case of structures having both rings and chains, the principal component must also have the most characteristic groups.

Examples

14.

1-(2-hydroxyethyl)cycloheptane-1,2-diol

15.

3-(4-hydroxycyclohexyl)propane-1,2-diol

In the names of amines, the general use of suffixes and prefixes is not always observed. Normally, the suffix -amine would be added to the name of the parent hydride and engender names such as methanamine (CH_3-NH_2). Further substitution on the nitrogen atom would then be indicated by prefixes, leading to names that appear very cumbersome, such as *N*-methylmethanamine for $(CH_3)_2NH$ and *N*,*N*-dimethylmethanamine for $(CH_3)_3N$. The traditional names of methylamine, dimethylamine and trimethylamine are much simpler. In these names, the term amine is not a suffix. It is, in fact, the name of the parent hydride, NH_3, which now serves as the basis of substitutive names. Names such as diethylamine and tributylamine are representative of the preferred nomenclature. Diamines are named accordingly, as with ethane-1,2-diamine for H_2N-CH_2-CH_2-NH_2 and propane-1,3-diamine for H_2N-$[CH_2]_3$-NH_2. There are allowed alternatives for these last two compounds: ethylenediamine and propane-1,3-diyldiamine.

Mancude ketones are also of special interest, as the =O group can only be accommodated on the carbon framework if there is a -CH_2- to replace in the parent hydride. For a mancude hydride, it is necessary to consider first that a double or triple bond is saturated, and then that >CH_2 becomes CO. This leaves an extra H on a neighbouring carbon. Take naphthalene as an example; the name 1,2-dihydronaphthalen-2-one accurately describes the two necessary operations that are needed: saturation and substitution. A quicker way, leading to a simpler name, is to consider that the saturation and the substitution are concomitant, with the result that only one hydrogen atom has been added to the cyclic system. This additional hydrogen atom is called an added hydrogen and it is represented in the name by the symbol *H* together with the appropriate locant.

Examples

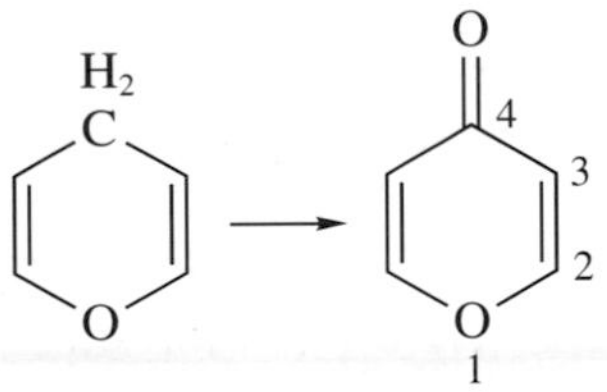

4*H*-pyran   pyran-4-one

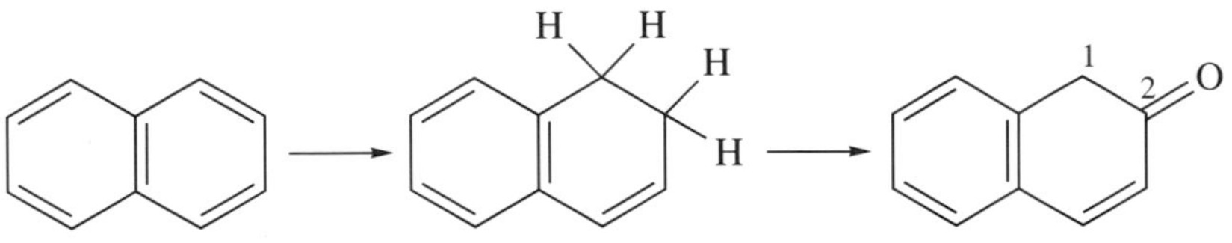

naphthalene   1,2-dihydronaphthalene   1,2-dihydronaphthalen-2-one
or naphthalen-2(1*H*)-one

The suffix -one is also used to name classes other than ketones. For instance, the lactones and the lactams, which are heterocyclic systems, can be named by adding the suffix -one to the name of the corresponding heterocycle. Specific suffixes -olactone and -olactam may also be used in simple cases.

Examples

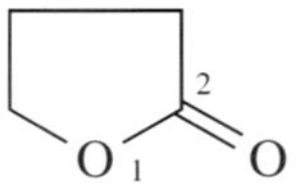

tetrahydrofuran-2-one or butano-4-lactone

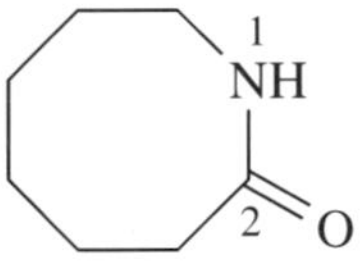

1-azocan-2-one or heptano-7-lactam

Two further rules are needed to name polyfunctional compounds:
1 There can be only one type of characteristic group, the principal group, which is cited as a suffix. The other groups must be cited as prefixes.
2 The principal group is selected using the priorities of Table 4.12.

Examples
20. CH_3-CO-CH_2-CH(OH)-CH_3 4-hydroxypentan-2-one
21. CH_3-C(O)-C(O)-C(O)OH   2,3-dioxobutanoic acid
22. H_2N-CH_2-CH_2-OH   2-aminoethanol
23. OHC-CH(CH_3)-CH_2-CH_2-COOH   4-methyl-5-oxopentanoic acid

24. OHC——⬡——COOH   4-formylcyclohexane-1-carboxylic acid

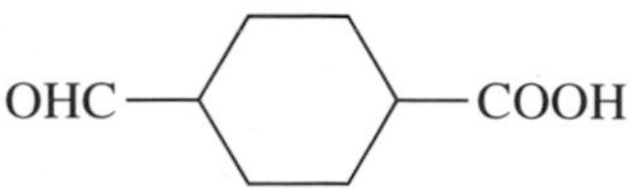

25. $H_2N\text{-}CO\text{-}CH_2\text{-}CH_2\text{-}COOH$ 4-amino-4-oxobutanoic acid

26.   2-(aminocarbonyl) benzoic acid

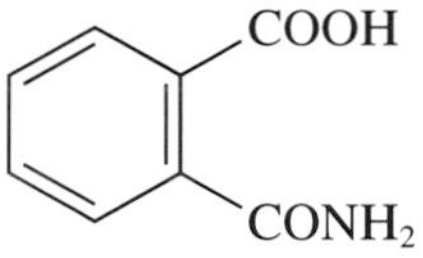

27. $NC\text{-}CH_2\text{-}CH_2\text{-}CH_2\text{-}CONH_2$ 4-cyanobutanamide

28.   3-cyanocyclobutane-1-carboxylic acid

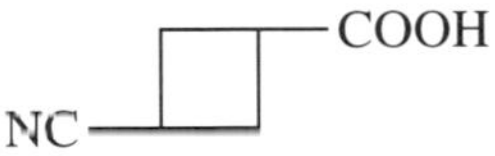

4.5.7   Names of functional parent hydrides

Parent hydrides are alkanes and mancudes having trivial or systematic names, and groups derived from these parent hydrides are indicated by non-detachable prefixes. However, there is a group of functional parent hydrides that is still known under trivial names. Acetic acid is an example. These functional parent hydride names are used like ordinary cyclic and acyclic parent hydride names, with one important difference. As they already contain characteristic groups prioritised to be cited as suffixes, they can be further functionalised only by characteristic groups having lesser priority, which will then be cited as prefixes. There are very few such functional parent hydride names recognised, but the names of some are given here due to their importance in nomenclature.

Hydrocarbons
$HC{\equiv}CH$ acetylene

Hydroxy compounds
C_6H_5OH phenol

Carbonyl compounds
$CH_3\text{-}CO\text{-}CH_3$ acetone

Carboxylic acids
$CH_3\text{-}COOH$ acetic acid
$CH_2{=}CH\text{-}COOH$ acrylic acid
$HOOC\text{-}CH_2\text{-}COOH$ malonic acid
$HOOC\text{-}CH_2\text{-}CH_2\text{-}COOH$ succinic acid
$H_2N\text{-}COOH$ benzoic acid
$C_6H_5\text{-}COOH$ oxamic acid
$H_2N\text{-}CO\text{-}COOH$ carbamic acid
$HOOC\text{-}COOH$ oxalic acid

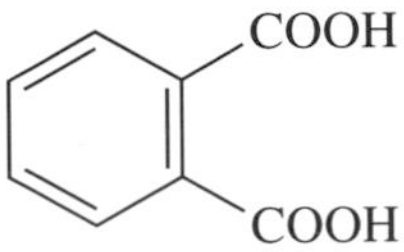

      phthalic acid

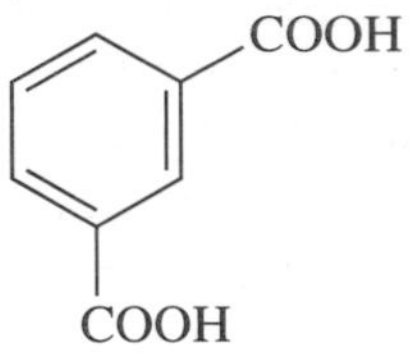

isophthalic acid

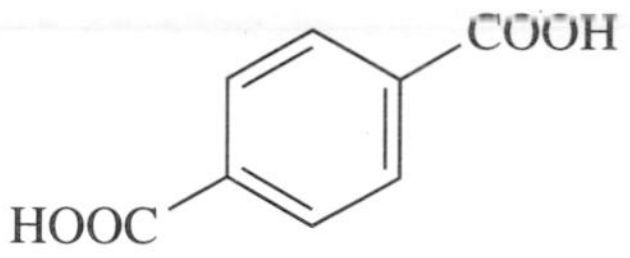

terephthalic acid

Amines

C$_6$H$_5$-NH$_2$ aniline

Acyclic polynitrogen compounds

H$_2$N-C(=NH)-NH$_2$ guanidine
H$_2$N-NH$_2$ hydrazine
H$_2$N-CO-NH$_2$ urea

Ring substituted benzenes, substitution in ring only

C$_6$H$_5$-CH$_3$ toluene
C$_6$H$_5$-CH=CH$_2$ styrene

Other names are retained for referring to unsubstituted compounds only. Compounds derived from them by substitution must be named systematically. The names are retained because of their wide use in biochemical and in polymer nomenclature. A few examples are given here.

Hydroxy compounds

HO-CH$_2$-CH$_2$-OH   ethylene glycol
HO-CH$_2$-CH(OH)-CH$_2$-OH   glycerol

Carboxylic acids

CH$_3$-CH$_2$-COOH   propionic acid
CH$_3$-CH$_2$-CH$_2$-COOH   butyric acid
HOOC-[CH$_2$]$_3$-COOH   glutaric acid
HOOC-[CH$_2$]$_4$-COOH   adipic acid
H$_2$C=C(CH$_3$)-COOH   methacrylic acid
H-COOH   formic acid (for nomenclature purposes, the hydrogen atom linked to carbon is not regarded as substitutable)

Amines

H$_3$C-C$_6$H$_4$-NH$_2$ toluidine (1,2-, 1,3- and 1,4- (*o*-, *m*- and *p*-) isomers)

The names propanoic acid (systematic) and propionic acid (retained) are both approved for the unsubstituted acid. However, the acid Cl-CH$_2$-CH$_2$-COOH must be named systematically as 3-chloropropanoic acid. The acid CH$_3$-CH(OH)-COOH is known as lactic acid, if unsubstituted; when it is substituted in position 3, for example with chlorine, the name becomes 3-chloro-2-hydroxypropanoic acid. Names such as 3-chloro-2-hydroxypropionic acid or 3-chlorolactic acid are not acceptable.

Retained names of carboxylic acids may also be modified to name amides, nitriles and aldehydes, by changing the -ic acid ending to -amide, -onitrile or -aldehyde. Names such as formaldehyde, acetonitrile and propionamide result. Of these, only acetonitrile may be treated as a functional parent hydride.

A more complete list of retained trivial names is shown in Table 4.13.

Table 4.13 Some trivial names still retained for naming organic compounds.

(A) Names of functional parent hydrides, usable with unlimited or limited substitution, as indicated

Hydrocarbons
$H_2C{=}C{=}CH_2$ — allene

Ethers
$C_6H_5\text{-}OCH_3$ — anisole (for ring substitution only, not methyl substitution)

Carbonyl compounds (ketones and pseudoketones)
$CH_2{=}C{=}O$ — ketene

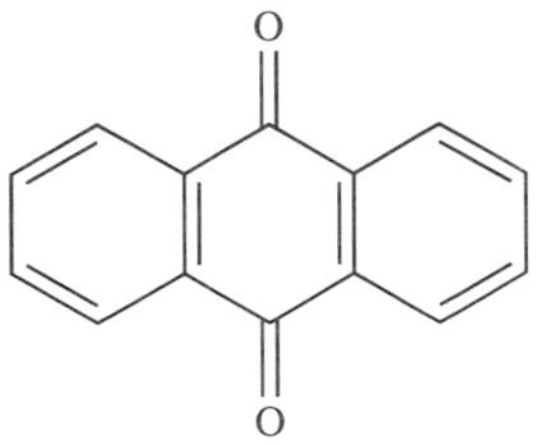

anthraquinone (9,10-isomer depicted)

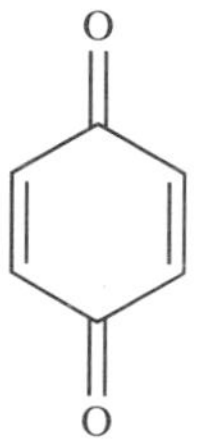

benzoquinone (1,4-isomer shown)

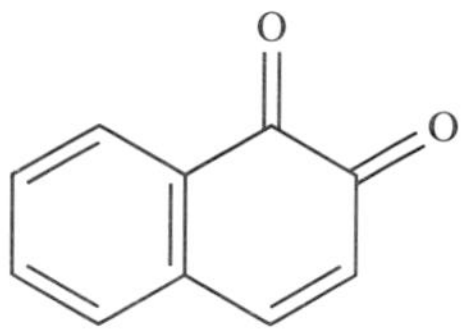

naphthoquinone (1,2-isomer shown)

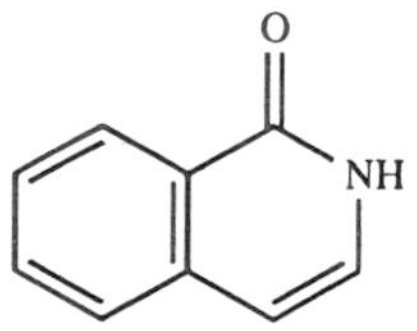

isoquinolone (1-isomer shown)

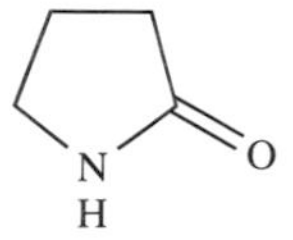

pyrrolidone (2-isomer shown)

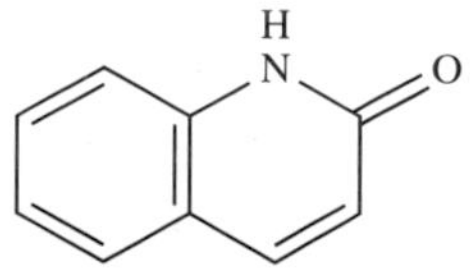

quinolone (2-isomer shown)

Continued on p. 94.

Table 4.13 (*Continued.*)

Carboxylic acids	
$CH_3\text{-}COOH$	acetic acid
$CH_2\!=\!CH\text{-}COOH$	acrylic acid
$HOOC\text{-}CH_2\text{-}COOH$	malonic acid
$HOOC\text{-}CH_2\text{-}CH_2\text{-}COOH$	succinic acid

(B) Names retained only for designating unsubstituted compounds

Hydroxy compounds	
$HO\text{-}CH_2\text{-}CH(OH)\text{-}CH_2\text{-}OH$	glycerol
$C(CH_2OH)_4$	pentaerythritol
$(CH_3)_2C\text{-}(OH)\text{-}C(OH)\text{-}C(CH_3)_2$	pinacol

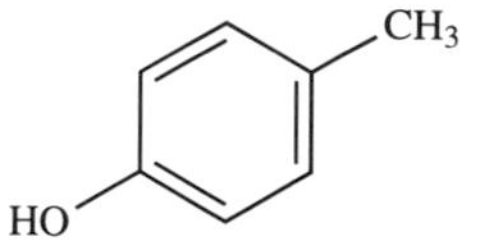

cresol (1,4-isomer shown)

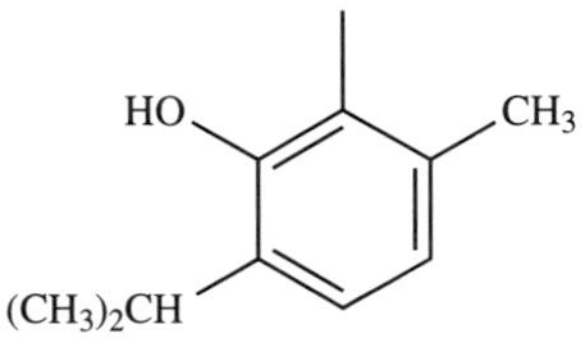

thymol

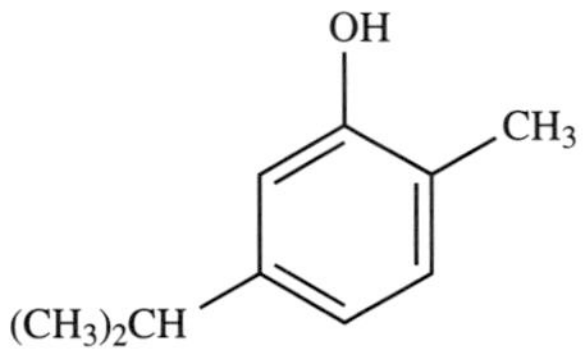

carvacrol

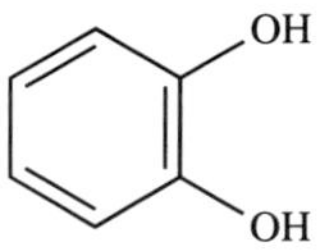

pyrocatechol

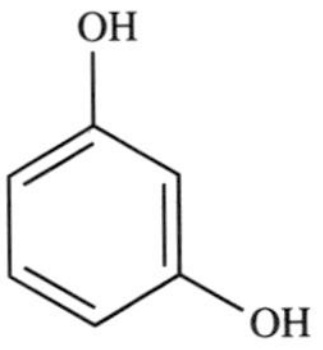

resorcinol

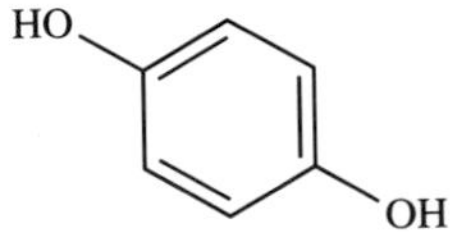

hydroquinone

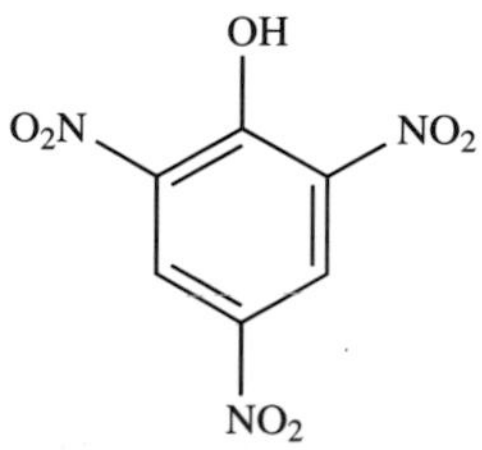

picric acid

Continued.

Table 4.13 (*Continued.*)

Carboxylic acids

CH_3-CH_2-COOH	propionic acid
CH_3-CH_2-CH_2-COOH	butyric acid
HOOC-$[CH_2]_3$-COOH	glutaric acid
HOOC-$[CH_2]_4$-COOH	adipic acid
H_2C=$C(CH_3)$-COOH	methacrylic acid
HOOC-$[CH(OH)]_2$-COOH	tartaric acid
$(HO)H_2C$-$CH(OH)$-COOH	glyceric acid
CH_3-$CH(OH)$-COOH	lactic acid
$(HO)H_2C$-COOH	glycolic acid
OHC-COOH	glyoxylic acid
CH_3-CO-CH_2-COOH	acetoacetic acid
CH_3-CO-COOH	pyruvic acid

CH_2-COOH
|
$C(OH)$-COOH   citric acid
|
CH_2-COOH

$(HOOC$-$CH_2)_2N$-CH_2-CH_2-$N(CH_2$-$COOH)_2$	ethylenediaminetetraacetic acid

Amines

H_3C-C_6H_4-NH_2	toluidine (1,2-, 1,3- and 1,4- (*o*-, *m*- and *p*-) isomers)

Hydrocarbons

H_2C=$C(CH_3)$-CH=CH_2	isoprene
H_3C-C_6H_4-CH_3	xylene (1,2-, 1,3- and 1,4- (*o*-, *m*- and *p*-) isomers)
$(CH_3)_2CH$-C_6H_5	cumene
H_3C-C_6H_4-$CH(CH_3)_2$	cymene (1,2-, 1,3- and 1,4- (*o*-, *m*- and *p*-) isomers)
1,3,5-$(CH_3)_3C_6H_3$	mesitylene

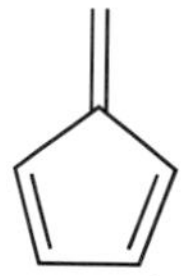

fulvene

4.5.8 Radicals and ions

Radicals and ions are not formed by a substitution operation, but by subtraction or addition of hydrogen atoms, hydrons or hydrides. Their names are formed using suffixes and prefixes, some of which are listed in Table 4.14.

Table 4.14 Some suffixes used to name radicals and ions.

	Operation	Suffix
Radicals	Loss of H·	-yl
	Loss of 2H·	-ylidene or -diyl*
Anions	Loss of H^+	-ide
Cations	Loss of H^-	-ylium
	Addition of H^+	-ium

* The suffix 'ylidene' is used to represent radicals such as R-CH: and/or R-ĊH, and the suffix 'diyl' when the electrons are localised on different atoms.

Examples

1. $CH_3^•$ methyl
2. CH_2: and/or $\overset{..}{C}H_2$ methylidene or carbene
3. $^•CH_2\text{-}CH_2^•$ ethane-1,2-diyl
4. $CH_3\text{-}CH_2^-$ ethanide or ethyl anion
5. CH_5^+ methanium
6. CH_3^+ methylium or methyl cation

These suffixes are cumulative, meaning that more than one can be present in a name, for instance to represent radical ions. Radicals come before anions, which in turn come before cations, in the order of seniority for citation as suffixes (cf. Table 4.12).

Examples

7. $CH_2^{•+}$ methyliumyl
8. $CH_4^{•+}$ methaniumyl
9. $^+CH_4\text{-}CH_2^•$ ethan-2-ium-1-yl
10. $^-CH_2\text{-}CH_2^•$ ethan-2-id-1-yl

The terms radical cation or radical ion(+) may be added to the name of the parent hydride when the positions of free valencies and/or charges are not known, or when it is not desirable to specify them.

Example

11. $(C_2H_6)^{•+}$ ethane radical cation or ethane radical ion(+)

4.5.9   **Stereochemical descriptors**

The use of stereochemical descriptors in both names and formulae is dealt with in Chapter 3, Section 3.8 (p. 21).

4.6   FUNCTIONAL CLASS NOMENCLATURE

Substitutive nomenclature is the nomenclature of choice in organic chemistry. However, it cannot be used to name all classes of compound. Salts, esters, acyl halides and anhydrides cannot be named substitutively when the characteristic group is chosen as the principal group, and functional class names (formerly called radicofunctional names) are then used. Functional class nomenclature is a binary system widely used in inorganic chemistry (see introduction to Chapter 4). Whereas substitutive names are generally written as one word, binary names are composed of two words. The names of salts of carboxylic acids are binary, as in sodium acetate. Names of esters and acyl halides are constructed in a similar way: methyl acetate for $CH_3\text{-}COOCH_3$, methyl chloroacetate for $ClCH_2\text{-}COOCH_3$, acetyl chloride for $CH_3\text{-}COCl$ and benzoyl bromide for $C_6H_5\text{-}COBr$. Prefixes for these groups are used in substitutive names when the principal group (cf. Table 4.12) is not an ester or an acyl halide.

Examples

1. ClOC-[CH$_2$]$_2$-COOH   4-chloro-4-oxobutanoic acid
2. C$_6$H$_5$-COO-[CH$_2$]$_3$-COOH   4-(benzoyloxy)butanoic acid

Functional modifiers are used in binary nomenclature to name anhydrides, e.g. acetic anhydride for (CH$_3$-CO)$_2$O, and derivatives of ketones and aldehydes such as oximes, hydrazones and semicarbazones, e.g. acetone oxime for (CH$_3$)$_2$C=NOH.

5 Aspects of the nomenclature of organometallic compounds

5.1 GENERAL

Organometallic compounds are defined as compounds containing a direct link between a carbon atom and a metal. What constitutes a metal for nomenclature purposes is really rather vague. The practice in nomenclature is to consider any element other than C, H and the rare gases to be metals if this is useful. The names of such compounds reflect their constitution and are drawn both from organic nomenclature and from inorganic nomenclatures. The names of organometallic compounds demonstrate that nomenclatures must be unified and adaptable to any situation.

Let us take the example of an organometallic compound derived from tin, e.g. $[Sn(C_2H_5)_4]$. Note that the square brackets imply a coordination compound. Using coordination nomenclature, its name is based on those of the central atom, tin, and of the ligands, ethyl. The resulting name, tetraethyltin, is constructed according to the principles of inorganic nomenclature (in being an additive name) and organic nomenclature (in being derived from a parent hydride, ethane). However, there is another approach to name this compound. The hydride of tin, SnH_4, is similar to that of carbon, CH_4 (methane). A name similar to methane can be coined for SnH_4 by attaching the ending 'ane' to the stem characteristic of tin, namely stann-. Stannane can be considered as a parent hydride and thus serve as the basis of substitutive names. The name tetraethylstannane is arrived at as a substitutive name.

Thus organometallic compounds can be named by an additive or a substitutive process. Additive nomenclature is applicable to all organometallic compounds, but substitutive nomenclature is arbitrarily restricted to names of derivatives of specific metals, the elements of Groups 14, 15, 16 and 17, and boron.

5.2 DERIVATIVES OF MAIN GROUP ELEMENTS

5.2.1 Selection of parent hydrides and their names

Carbon is, of course, unique in the number of hydrides it forms, but the elements in the proximity of carbon in the Periodic Table have a similar, if more restricted, propensity to form hydrides. Silicon, germanium, boron and phosphorus are obvious examples. For hydrides of these elements, and especially for their organic derivatives, the methods of substitutive nomenclature can be applied to obtain suitable names.

It is generally an arbitrary matter to decide where to apply substitutive nomenclature in these cases. Table 5.1 shows the elements to which both CNIC and CNOC approve the application. Table 5.2 gives the names of the corresponding mononuclear parent hydrides. The only additional elements to which substitutive nomenclature may sometimes be applied are the halogens, particularly iodine.

Table 5.1 Parent hydride elements.

Group 13	Group 14	Group 15	Group 16	Group 17	Group 18
B	C	N	O	F	Ne
Al	Si	P	S	Cl	Ar
Ga	Ge	As	Se	Br	Kr
In	Sn	Sb	Te	I	Xe
Tl	Pb	Bi	Po	At	Rn

Several points arise from Tables 5.1 and 5.2. The ending -ane signifies that the element exhibits its standard bonding number (i.e. the conventional number of electron-pair bonds), namely 3 for simple boron hydrides, 4 for the Group 14 elements, 3 for Group 15 elements and 2 for Group 16 elements. Where other bonding numbers are exhibited, this is indicated using the λ-convention (see the *Guide to IUPAC Nomenclature of Organic Compounds*, p. 21 and the *Nomenclature of Inorganic Chemistry*, p. 84). The alternative names appended are coordination names.

Examples

1. PH_3, phosphane or trihydridophosphorus
2. PH_5, λ^5-phosphane or pentahydridophosphorus
3. SH_2, sulfane or dihydridosulfur
4. SH_6, λ^6-sulfane or hexahydridosulfur

Several of the names for the parent hydrides, although systematic, are not in general use, and alternatives are approved: these are azane (ammonia is sanctioned by wide usage), oxidane (water) and sulfane (hydrogen sulfide).

The names of polynuclear hydrides (i.e. compounds with molecules consisting of chains) are obtained by prefixing the -ane names of Table 5.2 with the appropriate multiplicative prefixes of di-, tri-, tetra-, etc.

Table 5.2 Mononuclear parent hydrides.

BH_3	borane	NH_3	azane*	OH_2	oxidane*,†
SiH_4	silane	PH_3	phosphane*	SH_2	sulfane*
GeH_4	germane†	AsH_3	arsane*	SeH_2	selane†
SnH_4	stannane	SbH_3	stibane*	TeH_2	tellane†
PbH_4	plumbane	BiH_3	bismuthane†	PoH_2	polane

* Phosphine, arsine and stibine may be retained for the unsubstituted mononuclear hydrides and for use as derived ligands and in forming certain derived groups, but they are not recommended for naming substituted derivatives. The systematic names in substitutive nomenclature for ammonia (NH_3) and water (H_2O) are azane and oxidane, respectively. These names are usually not used, but are available if required. Sulfane, when unsubstituted, is usually named hydrogen sulfide. The normal formulae H_2O, H_2S, H_2Se, H_2Te and H_2Po have been reversed for purposes of comparison.
† Names based on such other forms as oxane, germanane, selenane, tellurane and bismane cannot be used because they are used as names for saturated six-membered heteromonocyclic rings based on the Hantzsch–Widman system.

Examples

	Substitutive names	Coordination names
5. H_2PPH_2	diphosphane	tetrahydridodiphosphorus($P-P$)
6. H_3SnSnH_3	distannane	hexahydridoditin($Sn-Sn$)
7. $SiH_3SiH_2SiH_2SiH_3$	tetrasilane	decahydridotetrasilicon($3Si-Si$)

The names of unsaturated compounds are derived by using appropriate substitutive nomenclature rules. Note that trivial names are also allowed for particular polynuclear species, for example, N_2H_4, diazane, commonly known as hydrazine. For a discussion of names of hydrides in which elements exhibit non-standard bonding numbers, see the *Nomenclature of Inorganic Chemistry*, p. 85. Note that for the hydrides of Table 5.1 and their derivatives, substitutive names are generally preferred.

5.2.2   **Names of substituted derivatives**

The preferred names are also obtained by applying the principles of substitutive nomenclature. Substituents, considered as replacing hydrogen atoms, are named using prefixes of the appropriate group names and are cited, if there is more than one, in alphabetical order before the name of the parent hydride, using parentheses and multiplicative prefixes as necessary.

Examples

1. $PH_2(CH_2CH_3)$	ethylphosphane
2. $Sb(CH=CH_2)_3$	trivinylstibane
3. $Si(OCH_2CH_2CH_3)Cl_3$	trichloro(propoxy)silane
4. $(C_2H_5)_3PbPb(C_2H_5)_3$	hexaethyldiplumbane
5. $C_3H_7SnH_2SnCl_2SnH_2Br$	1-bromo-2,2-dichloro-3-propyl-tristannane
6. $GeH(SCH_3)_3$	tris(methylsulfanyl)germane

Where it is not obvious which atom should be selected central atom, the choice may be made as indicated in the *Nomenclature of Inorganic Chemistry*, p. 87: P, As, Sb, Bi, Si, Ge, Sn, Pb, B, S, Se, Te, C.

Examples

7. $H_3CPHSiH_3$	methyl(silyl)phosphane
8. $Ge(C_6H_5)Cl_2(SiCl_3)$	trichloro[dichloro(phenyl)germyl]silane

Coordination names can be proposed for all these species, for example hexaethyldiplumbane may also be named hexaethyldilead($Pb-Pb$) or bis(trimethyllead)($Pb-Pb$). There seems no obvious advantage in the coordination names, and the substitutive names are usually used.

These examples also demonstrate how the -ane name of the parent hydride is adapted to give the name of the corresponding substituent group, as the name methane gives rise to the name methanyl, generally contracted to methyl. In these cases, silane gives rise to silyl, germane to germyl, etc., but phosphane to phosphanyl. The problem is quite complex because, for example, disilyl means $(SiH_3)_2$ whereas disilanyl means SiH_3SiH_2-. The reader is referred to the *Nomenclature of Inorganic Chemistry*, Chapter 7, for more detailed discussion. Finally, it should be noted that certain elements, such as sulfur and nitrogen, have well-developed alternative systems of nomenclature based upon other organic nomenclature principles. The reader is referred to the *Nomenclature of Organic Chemistry* for details.

5.2.3   **Skeletal replacement**

In some cases, it may be preferable to consider a parent hydride in which a skeletal carbon atom (or atoms) of a hydrocarbon has (have) been replaced by some other atom(s). The following example demonstrates the principle, even if it is too simple to typify general practice. Thus, $Si(CH_2CH_3)_2H_2$ is probably best named diethylsilane. However, if written in the form CH_3-CH_2-SiH_2-CH_2-CH_3, it can be seen to be a derivative of pentane in which the central carbon atom has been replaced by a silicon atom. Using skeletal replacement methodology, this can also be named 3-sila-pentane, although this is probably not as useful as the silane name. In general, such names should be used only when four or more heteroatoms are present.

The modified element name 'sila' indicates replacement in the carbon skeleton, and similar treatment can be applied to other element names. The parent hydride names of Table 5.2 may all be adapted in this way and used in the same fashion as in the oxa-aza nomenclature of organic chemistry. In inorganic chemistry, a major use is in names of cyclic derivatives that have heteroelement atoms replacing carbon atoms in structures. It may be possible to name such species by Hantzsch–Widman procedures (see p. 77), and these should always be used when applicable.

The reader is referred to the *Nomenclature of Inorganic Chemistry* and to the *Nomenclature of Organic Chemistry* for detailed treatments. The application can be complex, but on occasion the usage can be advantageous.

Examples

1. 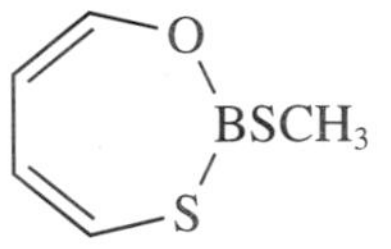   2-(methylsulfanyl)-1,3,2-oxathiaborepine

2. 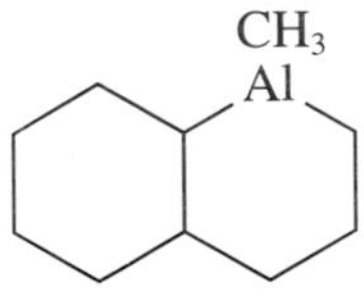   1-methyldecahydro-1-aluminanaphthalene

3. 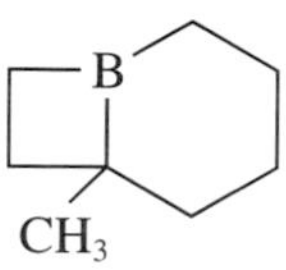   6-methyl-1-borabicyclo[4.2.0]octane

4. 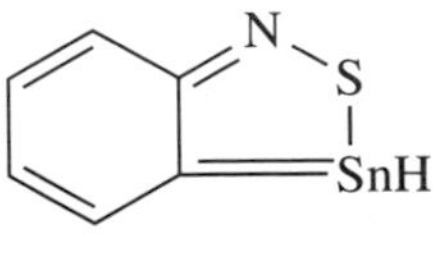   2,1,3-benzothiazastannole

5.3 ORGANOMETALLIC DERIVATIVES OF
TRANSITION ELEMENTS

For these derivatives, coordination nomenclature is generally preferred. The procedures and devices have been dealt with in Chapter 4, Section 4.4 (p. 51), and the reader is referred there and to the *Nomenclature of Inorganic Chemistry* for more details.

6 Macromolecular (polymer) nomenclature

6.1 DEFINITIONS

This discussion is based on the 1991 edition of the *Compendium of Macromolecular Nomenclature* and several more recent recommendations of the IUPAC Commission on Macromolecular Nomenclature.[1]

- A macromolecule, or polymer molecule, is a molecule of high relative molecular mass, the structure of which essentially comprises the multiple repetitions of units derived from molecules of low relative mass.

- A monomer is a substance composed of monomer molecules that can undergo polymerisation, thereby contributing constitutional units to the essential structure of a macromolecule. A constitutional unit is an atom or group of atoms (with pendant atoms or groups, if any) comprising a part of the essential structure of a macromolecule.

- Polymerisation is the process of converting a monomer, or a mixture of monomers, into a polymer.

- A monomeric unit is the largest constitutional unit contributed by a single monomer molecule to the structure of a macromolecule.

- A constitutional repeating unit is the smallest constitutional unit, the repetition of which constitutes a regular macromolecule.

- A regular polymer is a substance composed of regular macromolecules, the structure of which essentially comprises the repetition of a single constitutional unit with all units connected identically with respect to directional sense.

- An irregular polymer is a substance composed of irregular macromolecules, the structure of which essentially comprises the repetition of more than one type of constitutional unit, or of macromolecules, the structure of which comprises constitutional units not all connected identically with respect to directional sense.

- A single-strand polymer is composed of single-strand macromolecules, the structure of which comprises constitutional units connected in such a way that adjacent constitutional units are joined to each other through two atoms, one on each constitutional unit.

- A double-strand polymer is composed of double-strand macromolecules, the structure of which comprises constitutional units connected in such a way that adjacent constitutional units are joined to each other through three or four atoms, two on one side and either one or two on the other side of each constitutional unit.

1. IUPAC, Commission on Macromolecular Nomenclature, *Compendium of Macromolecular Nomenclature*. Blackwell Scientific Publications, Oxford (1991).
IUPAC, Commission on Macromolecular Nomenclature, Nomenclature of regular double-strand (ladder and spiro) organic polymers. *Pure Appl. Chem.*, **65**, 1561–1580 (1993).
IUPAC, Commission on Macromolecular Nomenclature, Structure-based nomenclature for irregular single-strand polymers. *Pure Appl. Chem.*, **66**, 873–879 (1994).
IUPAC, Commission on Macromolecular Nomenclature, Graphic representations (chemical formulae) of macromolecules. *Pure Appl. Chem.*, **66**, 2469–2482 (1994).
IUPAC, Commission on Macromolecular Nomenclature, Glossary of basic terms in polymer science. *Pure Appl. Chem.*, **68**, 2287–2311 (1996).

Examples of a double-strand polymer are a ladder polymer in which macromolecules consist of an uninterrupted sequence of rings with adjacent rings having two or more atoms in common, and a spiro polymer in which macromolecules consist of an uninterrupted sequence of rings, with adjacent rings having only one atom in common.

• A homopolymer is a polymer derived from one species of monomer.

• A copolymer is a polymer derived from more than one species of monomer.

• A block copolymer is composed of block macromolecules in which adjacent linear blocks are constitutionally different, i.e. adjacent blocks comprise constitutional units derived from different species of monomers or from the same species of monomer but with different composition or sequence distribution of constitutional units.

• A graft copolymer is composed of graft macromolecules with one or more species of block connected to the main chain as side-chains, these side-chains having constitutional or configurational features that differ from those in the main chain.

• An end-group is a constitutional unit that is an extremity of a macromolecule.

6.2   GENERAL CONSIDERATIONS

Polymers are unlike low-molecular-weight compounds in that they have no uniform structure and are a mixture of macromolecules of different length and different structural arrangements, even when derived from a single monomer.

For instance, poly(vinyl chloride), derived from the polymerisation of vinyl chloride (chloroethene), $CH_2{=}CHCl$, contains repeating units $-CH_2-CHCl-$. However, long-chain molecules are of various lengths and the units are not necessarily all uniquely oriented and joined in a regular fashion, which would result in the polymer formulated: $-(CH_2-CHCl)_{\overline{n}}$.

In addition to 'head-to-tail' links, $-CH_2-CHCl-CH_2-CHCl-$, other links such as 'head-to-head' or 'tail-to-tail' can occur, as in $-CH_2-CHCl-CHCl-CH_2-CH_2-CHCl-$.

This becomes even more complicated in the structure of a copolymer derived from more than one species of monomer, such as styrene and methyl acrylate, both of which contribute constitutional units.

Examples

These can combine in turn in a polymeric chain in a variety of ways, resulting in types such as unspecified, statistical, random, alternating, periodic, block and graft copolymers.

Because the exact structure of the polymer is not always known, two systems of macromolecular nomenclature have been developed: source-based nomenclature and structure-based nomenclature.

6.3   SOURCE-BASED NOMENCLATURE

The name of the polymer is formed by attaching the prefix 'poly' to the name of the
real or assumed monomer, or the starting material (source) from which the polymer
is derived.

Examples
1. polyacrylonitrile
2. polyethylene
3. poly(methyl methacrylate)
4. polystyrene
5. poly(vinyl acetate)
6. poly(vinyl chloride)

Parentheses are used when the name of the monomer consists of two or more words.
    For copolymers, a connective (infix) is inserted, which depends on what is known
about the arrangement of the constitutional units.

Examples
7. poly[styrene-*co*-(methyl methacrylate)] or copoly(styrene/methyl methacrylate)
   (an unknown or unspecified arrangement)
8. poly(styrene-*stat*-acrylonitrile-*stat*-butadiene) or *stat*-copoly(styrene/-
   acrylonitrile/butadiene) (a statistical arrangement, obeying known statistical
   laws)
9. poly[ethylene-*ran*-(vinyl acetate)] or *ran*-copoly(ethylene/vinyl acetate) (a
   random arrangement, with a Bernoullian distribution)
10. poly[(ethylene glycol)-*alt*-(terephthalic acid)] or *alt*-copoly(ethylene glycol/-
    terephthalic acid) (an alternating sequence)
11. poly[[(ethylene glycol)-*alt*-(terephthalic acid)]-*co*-[(ethylene glycol)-*alt*-
    (isophthalic acid)]] or *alt*-copoly[ethylene glycol/(terephthalic acid;
    isophthalic acid)] (an unspecified arrangement of two alternating pairs of
    monomers)
12. polystyrene-*block*-polybutadiene or *block*-copoly(styrene/butadiene) (a linear
    arrangement of blocks, such as -AAAA-BBBB-)
13. polybutadiene-*graft*-polystyrene or *graft*-copoly(butadiene/styrene) (a graft
    arrangement, such as −AAAA′AAA−
                                   |
                                   B
                                   B
                                   B
                                   |

6.4   STRUCTURE-BASED NOMENCLATURE

6.4.1   **Regular single-strand organic polymers**

For regular organic polymers that have only one species of constitutional repeating
unit (CRU) in a single sequential arrangement and consist of single strands, the name
is of the form poly(CRU).

The CRU is named as an organic divalent group according to the usual rules for organic chemistry. The steps involved are: identification of the unit; orientation of the unit; and naming of the unit.

For instance, in a polymer such as

$$-O-CH-CH_2-O-CH-CH_2-O-CH-CH_2-O-$$
$$\quad\ \ |\qquad\qquad\quad |\qquad\qquad\quad |$$
$$\quad\ \ Cl\qquad\qquad\quad Cl\qquad\qquad\quad Cl$$

the CRU may be identified in at least three ways:

$$-O-CH-CH_2-\qquad\qquad -O-CH_2-CH-\qquad\qquad -CH_2-O-CH-$$
$$\quad\ \ |\qquad\quad or\qquad\qquad\qquad |\qquad\quad or\qquad\qquad\qquad |$$
$$\quad\ \ Cl\qquad\qquad\qquad\qquad\qquad Cl\qquad\qquad\qquad\qquad\qquad Cl$$

To obtain a unique name, a single preferred CRU must be selected. Therefore, rules have been developed that specify both seniority among subunits, that is, the point at which to begin writing the CRU, and also the direction in which to move along the chain to reach the end of the CRU.

The order of seniority among the types of divalent group is:

1 heterocyclic rings;
2 chains containing heteroatoms;
3 carbocyclic rings;
4 chains containing only carbon.

Within each structural type, the seniority is dictated by the seniority of individual constituents. Examples of some of the rules are given below:

• For heterocyclic rings, a ring system containing nitrogen is senior to a ring system containing a heteroatom other than nitrogen, with further descending order of seniority governed by the greatest number of rings in a ring system, the largest

Table 6.1 Seniority of ring systems

The seniority of ring systems is decided by applying the following criteria, successively in the order given, until a single ring is left under consideration.
(a) All heterocycles are senior to all carbocycles.
(b) For heterocycles, the order of seniority is:
    1 a nitrogen-containing ring;
    2 a ring containing another different heteroatom as high as possible in Table 4.8;
    3 a system containing the greatest number of rings;
    4 a system containing the largest possible individual ring;
    5 a system containing the greatest number of heteroatoms of any kind;
    6 a system containing the greatest variety of heteroatoms;
    7 a system containing the greatest number of heteroatoms listed earliest in Table 4.8;
    8 a fused system with the lowest locants for the heteroatom before fusion.
(c) Largest number of rings.
(d) Largest individual ring at first point of difference.
(e) Largest number of atoms in common among rings.
(f) Lowest letters (lowest means *a* lower than *b*, etc.) in the expression of ring junctions in fusion nomenclature.
(g) Lowest numbers at the first point of difference in the expression for ring junctions.
(h) Lowest state of hydrogenation.
(i) Lowest locant for indicated hydrogen.
(j) Lowest locant for points of attachment (if a substituent).
(k) Lowest locant for an attached group expressed as a suffix.
(l) Lowest locant for substituents named as prefixes.
(m) Lowest locant for that substituent named as prefix that is cited first in the name.

individual ring in a ring system, the largest number of heteroatoms and the greatest variety of heteroatoms (for details, see Table 6.1).

• For an acyclic chain containing a heteroatom, oxygen is senior to sulfur, sulfur to nitrogen, nitrogen to phosphorus, phosphorus to silicon, silicon to germanium, etc.

• For carbocyclic rings, a three-ring system is senior to a two-ring system, a two-ring system containing two six-membered rings is senior to one containing a five- and a six-membered ring, a fused two-ring system (two atoms common to both rings) is senior to a spiro ring system (one atom in common) of the same size and an unsaturated ring is senior to a saturated ring of the same size.

• For chains containing carbon, seniority is determined first by length, then by the number of substituents, then by the locants and, finally, by the alphabetical order of substituents.

If the CRU has identical subunits separated by other subunits, the direction of citation is determined by the shorter path between them.

The preferred CRU is the one beginning with the subunit of highest seniority. To establish direction, one proceeds from this subunit to the neighbouring subunit of the same or next in seniority. In the above example, the subunit of highest seniority is an oxygen atom and the subunit next in seniority is a substituted $-CH_2CH_2-$. The substituted subunit is oriented in such a way that the substituent, the chlorine atom, is assigned the lowest locant. The CRU is written to read from left to right. Thus, the preferred CRU is

$$-O-CH-CH_2-$$
$$\qquad |$$
$$\qquad Cl$$

and the polymer

$$+O-CH-CH_2+_n$$
$$\qquad |$$
$$\qquad Cl$$

is named poly[oxy(1-chloroethylene)].

The second example shows the CRU starting with a substituted nitrogen atom and proceeding through the shortest path to the unsubstituted nitrogen atom and then through a carbocycle.

poly[(methylimino)methyleneimino-1,3-phenylene]

The third example shows the CRU starting with a heterocyclic ring and proceeding through a substituted carbon atom to a heteroatom.

poly(pyridine-3,5-diylcarbonyloxymethylene)

Examples of some common polymers

	Structure-based name	Source-based or trivial name

1. $\left.\!\!-\!\!\left(CH\text{-}CH_2\right)\!\!-\!\!\right._n$ (with phenyl group)

    poly(1-phenylethylene)   polystyrene

2. $\left.\!\!-\!\!\left(\underset{CN}{CH}\text{-}CH_2\right)\!\!-\!\!\right._n$

    poly(1-cyanoethylene)   polyacrylonitrile

3. $\left.\!\!-\!\!\left(O\text{—}\bigcirc\right)\!\!-\!\!\right._n$

    poly(oxy-1,4-phenylene)   poly(phenylene oxide)

4. $\left.\!\!-\!\!\left(O\text{-}CH_2\text{-}CH_2\text{-}O\text{-}\underset{O}{\overset{\parallel}{C}}\text{—}\bigcirc\text{—}\underset{O}{\overset{\parallel}{C}}\right)\!\!-\!\!\right._n$

    poly(oxyethylene-oxyterephthaloyl)   poly(ethylene terephthalate)

5. $\left.\!\!-\!\!\left(NH\text{-}\underset{O}{\overset{\parallel}{C}}\text{-}[CH_2]_4\text{-}\underset{O}{\overset{\parallel}{C}}\text{-}NH\text{-}[CH_2]_6\right)\!\!-\!\!\right._n$

    poly(iminoadipoyl-iminohexamethylene)   poly(hexamethylene adipamide)

6. $\left.\!\!-\!\!\left(\underset{O=C\underset{\underset{O}{}}{}C=O}{CH\text{-}CH}\text{———}CH\text{-}CH_2\right)\!\!-\!\!\right._n$ (with phenyl group)

    poly[(2,5-dioxotetrahydrofuran-3,4-diyl)(1-phenylethylene)]   poly(maleic anhydride-*alt*-styrene)

If the end-groups of the chain are known, they may be specified by prefixes to the name of the polymer, with the symbol α designating the beginning or left-hand end-group and the symbol ω designating the other end-group.

Example

7. $Cl_3C\left.\!\!-\!\!\left(\bigcirc\text{-}CH_2\right)\!\!-\!\!\right._n Cl$

    α-(trichloromethyl)-ω-chloropoly(1,4-phenylenemethylene)

6.4.2   Regular double-strand organic polymers

In a double-strand polymer, the macromolecules consist of an uninterrupted sequence of rings with adjacent rings having one atom in common (a spiro polymer) or two or more atoms in common (a ladder polymer).

As for a single-strand polymer, a single preferred CRU must be selected in order to obtain a unique name. The CRU is usually a tetravalent group denoting attachment to four atoms and is named according to the usual rules of organic nomenclature. The name of the polymer is in the form: poly(CRU).

Because the polymer has a sequence of rings, in order to identify a preferred

CRU, the rings must be broken by observing the following criteria in decreasing order of priority:

1 minimise the number of free valencies in the CRU;

2 maximise the number of most-preferred heteroatoms in the ring system;

3 retain the most preferred ring system;

4 choose the longest chain for acyclic CRU.

Further decisions are based on the seniority of ring systems (for details, see Table 6.1), on the orientation of the CRU to give the lowest free valence locant at the lower left of the structural diagram and on placing the acyclic subunits, if any, on the right side of the ring system within the CRU.

For a polymer consisting of adjacent six-membered saturated carbon rings

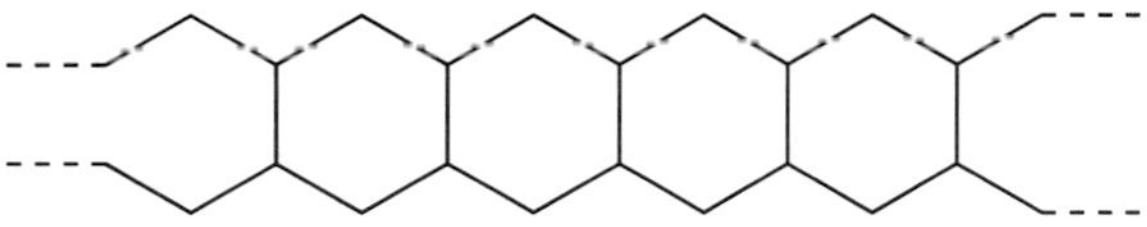

the name based on the preferred CRU is

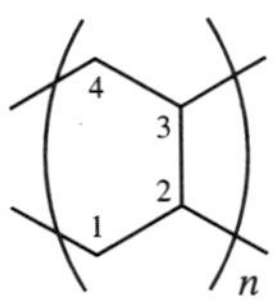

poly(butane-1,4:3,2-tetrayl)

For a polymer consisting of alternating six-membered sulfur-containing and keto-group-containing rings

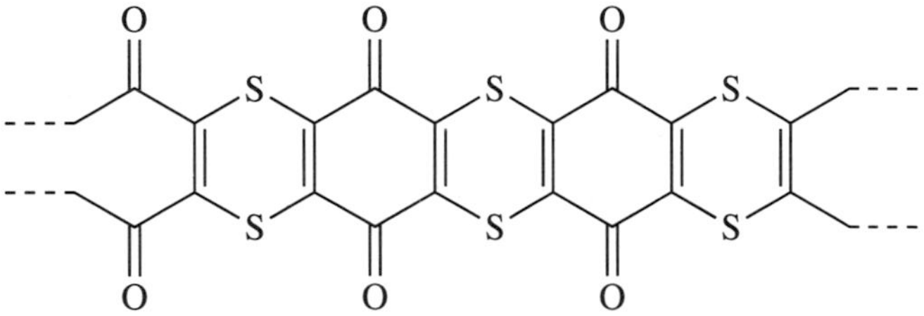

the name based on the preferred CRU is

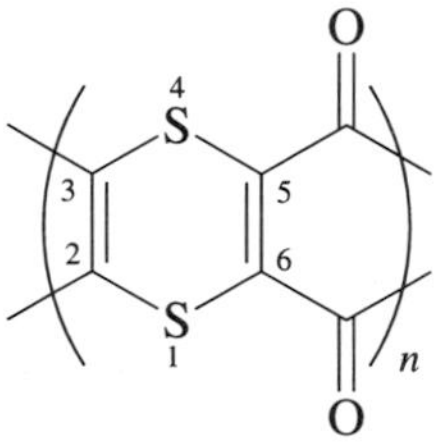

poly(1,4-dithiin-2,3:5,6-tetrayl-5,6-dicarbonyl)

For a benzobisimidazobenzophenanthroline-type (BBL) ladder polymer derived from 1,4,5,8-naphthalenetetracarboxylic acid and 1,2,4,5-benzenetetramine the name is poly[(7-oxo-7H,10H-benz[de]imidazo[4',5':5,6]benzimidazo-[2,1-a]isoquinoline-3,4:10,11-tetrayl)-10-carbonyl].

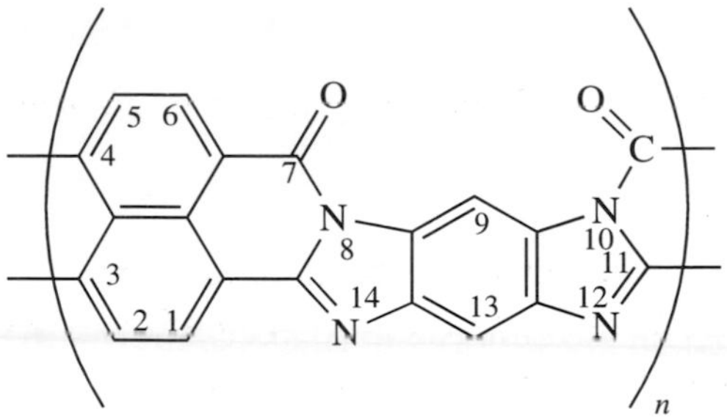

For a polymer consisting of adjacent cyclohexane and 1,3-dioxane rings in a regular spiro sequence, such as

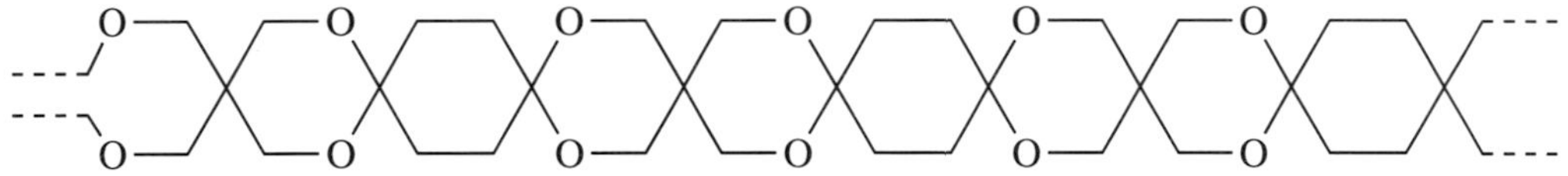

the name based on the preferred CRU is

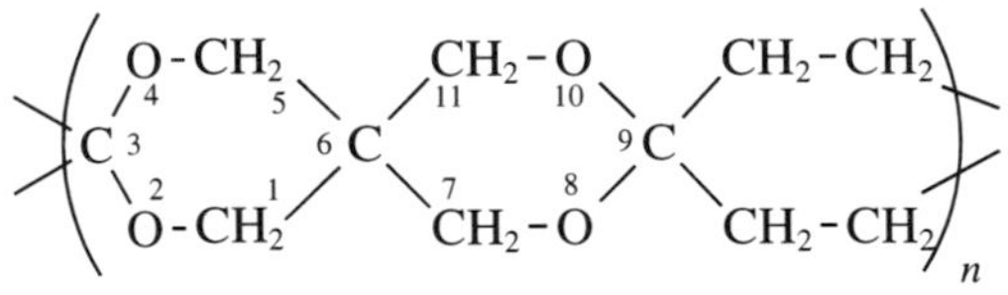

poly(2,4,8,10-tetraoxaspiro[5.5]undecane-3,3:9,9-tetrayl-9,9-diethylene)

6.4.3   Regular single-strand inorganic and coordination polymers

The names of inorganic and coordination polymers are based on the same fundamental principles that were developed for organic polymers. As in the nomenclature of organic polymers, these rules apply to structural representations that may at times be idealised and do not take into account irregularities, chain imperfections or random branching.

A CRU is selected and named. The name of the polymer is the name of the CRU prefixed by the terms 'poly', '*catena*' or other structural indicator. In order to arrive at the preferred CRU, the seniorities of the constituent subunits are considered, as well as the preferred direction for the sequential citation.

The constituent subunit of highest seniority must contain one or more central atoms; bridging groups between central atoms in the backbone of the polymer cannot be senior subunits. This is consistent with the principle of coordination nomenclature, in which the emphasis is laid on the coordination centre.

Examples

1. 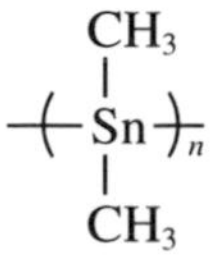   *catena*-poly[dimethyltin]

2. 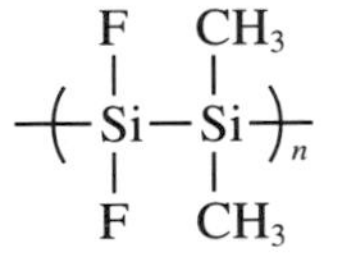 $\underset{\underset{F}{|}\quad \underset{CH_3}{|}}{\overset{\overset{F}{|}\quad \overset{CH_3}{|}}{\left(\!-Si-Si-\!\right)_n}}$ *catena*-poly[(difluorosilicon)(dimethylsilicon)]

The above examples contain homoatomic backbones. However, coordination polymers commonly consist of a single central atom with a bridging ligand. Such a polymer is named by citing the central atom prefixed by its associated non-bridging ligands, followed by the name of the bridging ligand prefixed by the Greek letter μ.

Examples

3. 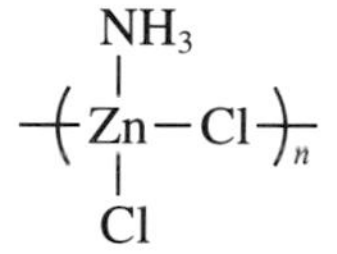 $\underset{\underset{Cl}{|}}{\overset{\overset{NH_3}{|}}{\left(\!-Zn-Cl-\!\right)_n}}$ *catena*-poly[(amminechlorozinc)-μ-chloro]

4. 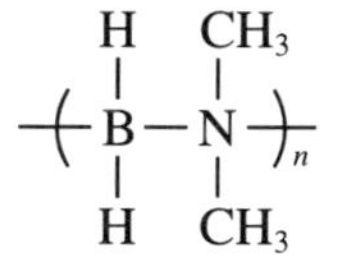 $\underset{\underset{H}{|}\quad \underset{CH_3}{|}}{\overset{\overset{H}{|}\quad \overset{CH_3}{|}}{\left(\!-B-N-\!\right)_n}}$ *catena*-poly[(dihydroboron)-μ-(dimethylamido)]

Multiple bridging ligands between the pair of central atoms are cited in alphabetical order. Italicised element symbols indicating the coordinating atoms in bridging ligands are cited in the order of direction of the CRU and are separated by a colon.

Example

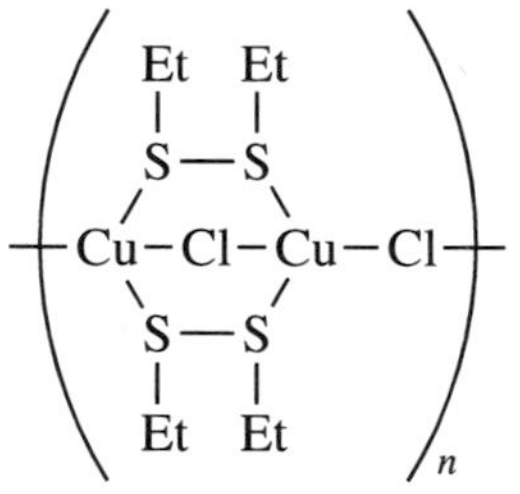

catena-poly[copper-[μ-chloro-bis-μ-(diethyl disulfide-*S*:*S*′)]-copper-μ-chloro]

6.4.4   Regular quasi-single-strand coordination polymers

A regular linear polymer that can be described by a preferred CRU in which *only one* terminal constituent subunit is connected through a single atom to the other identical CRU is a quasi-single-strand polymer. Such polymers are named similarly to single-strand coordination polymers.

Examples

1.   *catena*-poly[palladium-di-μ-chloro]

2. 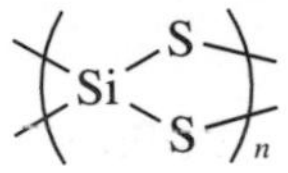 *catena*-poly[silicon-di-μ-thio]

3. 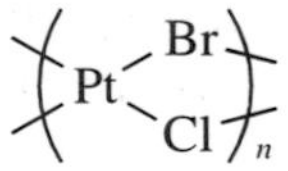 *catena*-poly[platinum(μ-bromo-μ-chloro)]

4. 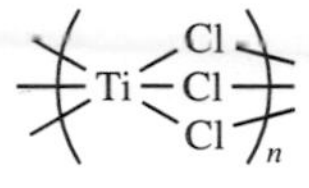 *catena*-poly[titanium-tri-μ-chloro]

6.4.5 Irregular single-strand organic polymers

Irregular polymers are named by placing the prefix 'poly' before the structure-based names of the constitutional units, collectively enclosed in parentheses or brackets, with the individual constitutional units separated by oblique strokes.

For instance, a partially hydrolysed poly(vinyl acetate) containing units

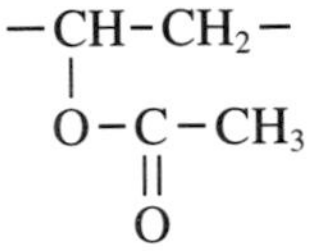

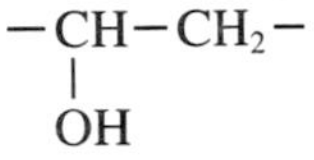

is named poly(1-acetoxyethylene/1-hydroxyethylene).

A copolymer of vinyl chloride and styrene consisting of units joined head-to-tail

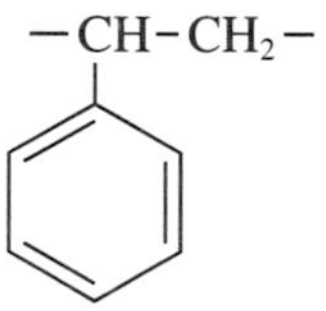

is named poly(1-chloroethylene/1-phenylethylene).

A chlorinated polyethylene consisting of units

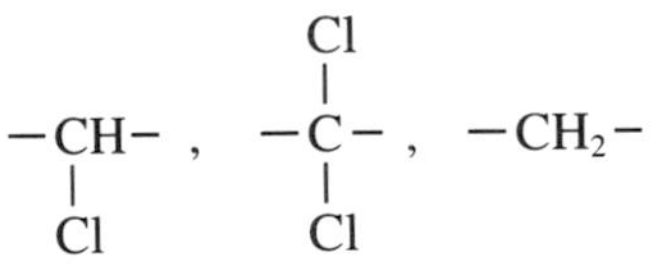

is named poly(chloromethylene/dichloromethylene/methylene).

A diblock copolymer in which the blocks of poly(ethylene oxide) and poly(vinyl chloride) are joined by a specific junction unit

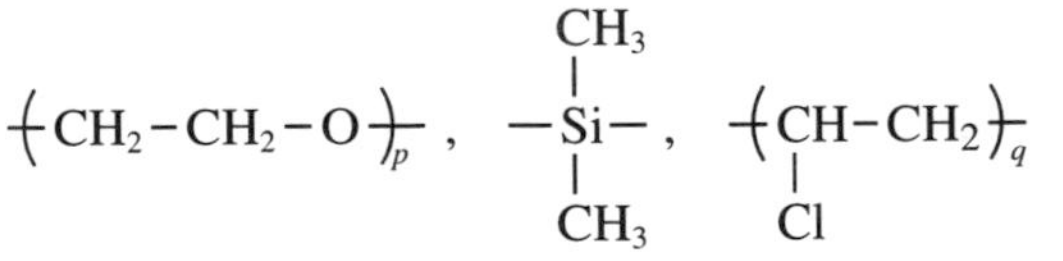

is named poly(ethyleneoxy)–dimethylsilanediyl–poly(1-chloroethylene).

6.5 TRADE NAMES AND ABBREVIATIONS

Trade names and abbreviations are frequently used in the literature and in oral communication, e.g. 'nylon 66' for poly(hexamethylene adipamide), 'Teflon' or PTFE for poly(tetrafluoroethylene) and 'Lucite' or PMMA for poly(methyl methacrylate). Other common abbreviations are listed below:

ABS	acrylonitrile/butadiene/styrene copolymer
PAN	polyacrylonitrile
PBT	poly(butylene terephthalate)
PEO	poly(ethylene oxide)
PET	poly(ethylene terephthalate)
PP	polypropylene
PS	polystyrene
PVAC	poly(vinyl acetate)
PVAL	poly(vinyl alcohol)
PVC	poly(vinyl chloride)
PVDF	poly(vinylidene difluoride)

7   Biochemical nomenclature

INTRODUCTION

Systematic substitutive nomenclature may be used to name all organic molecules. However, those that are of animal or vegetable origin have often received trivial names, such as cholesterol, oxytocin and glucose. Biochemical nomenclature is based upon such trivial names, which are either substitutively modified in accordance with the principles, rules and conventions described in Chapter 4, Section 4.5 (p. 70), or transformed and simplified into names of stereoparent hydrides, i.e. parent hydrides of a specific stereochemistry. These names are then modified by the rules of substitutive nomenclature. Three classes of compound will be discussed here to illustrate the basic approach: carbohydrates; amino acids and peptides; and lipids. For details, see: *Biochemical Nomenclature and Related Documents*, 2nd Edition, Portland Press, London (1992).

7.2   CARBOHYDRATE NOMENCLATURE

Originally, carbohydrates were defined as compounds such as aldoses and ketoses having the stoichiometric formula $C_n(H_2O)_n$, hence 'hydrates of carbon'. The generic term 'carbohydrates' includes monosaccharides, oligosaccharides and polysaccharides, as well as substances derived from monosaccharides by reduction of the carbonyl group (alditols), by oxidation of one or more terminal groups to carboxylic acid(s) or by replacement of one or more hydroxyl group(s) by a hydrogen atom, an amino group, thiol group or similar group. It also includes derivatives of these compounds. The term carbohydrate is synonymous with the term saccharide.

Trivial names are common in carbohydrate nomenclature. Fifteen of them form the basis of the systematic nomenclature. They are assigned to the simple aldoses (polyhydroxyaldehydes), from triose to hexoses.

Triose:	glyceraldehyde (*not* glycerose)
Tetroses:	erythrose, threose
Pentoses:	arabinose, lyxose, ribose, xylose
Hexoses:	allose, altrose, galactose, glucose, gulose, idose, mannose, talose

Among the keto-hexoses, fructose is of major natural occurrence.

A stereodescriptor, D or L, is used to indicate the absolute configuration of the entire molecule. Reference is made to the configuration of glyceraldehyde through a reference carbon atom, which is the carbon atom receiving the highest numerical locant (marked → in the examples below), the lowest possible locant being given to the carbon atom bearing the principal characteristic group, which is a carbonyl group in aldoses and ketoses.

Examples

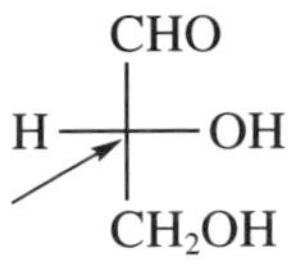

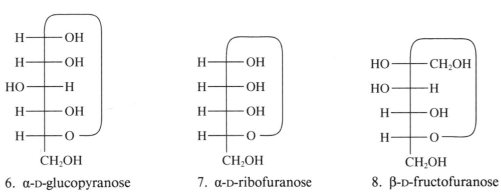

1. D glyceraldehyde 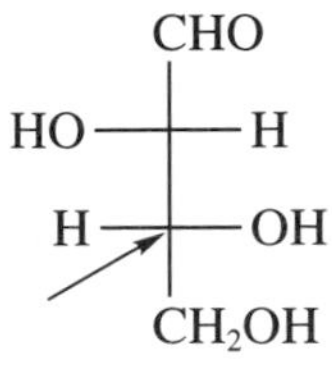   2. D-threose 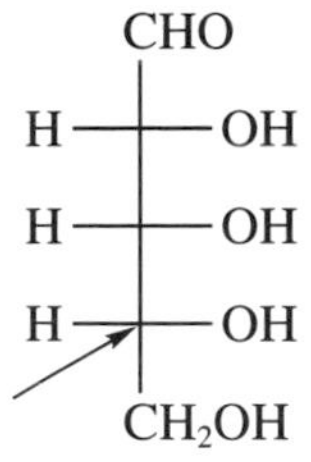   3. D-ribose 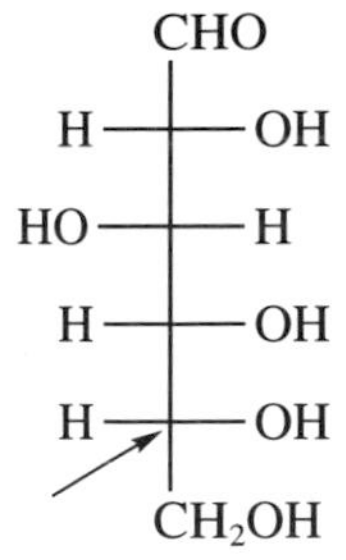   4. D-glucose 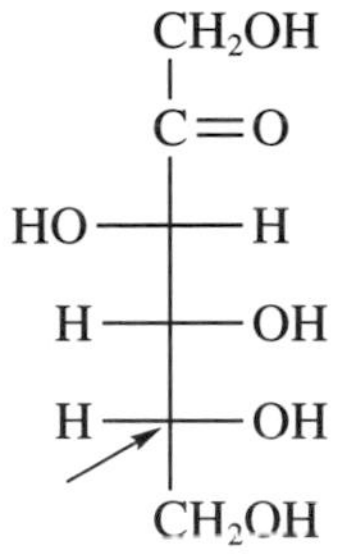   5. D-fructose

Names of cyclised (hemi-acetalised) aldoses and ketoses contain the infixes pyran or furan to indicate the six- or five membered heterocyclic structure and a stereo-descriptor, α or β, to indicate the configuration of the anomeric or hemi-acetal carbon atom.

Examples

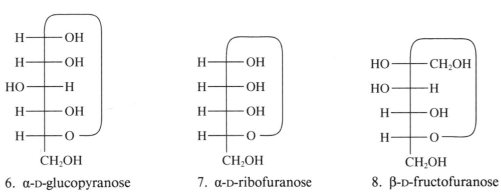

6. α-D-glucopyranose   7. α-D-ribofuranose   8. β-D-fructofuranose

Although the names of the saccharides are generally trivial, systematic nomenclature is used to name their derivatives. Because trivial names are not amenable to the treatments usually applied to the names of ordinary parent hydrides, many adaptations are necessary and some peculiarities must be noted.

For example, substitution can be made on an oxygen atom in the case of esters and ethers. It is characterised by the symbol *O*, which is placed after the locant. The compound prefix deoxy- is composed of the prefixes de-, meaning 'without' in subtractive nomenclature, and oxy-, to indicate the subtraction of an oxy group from an -OH group: C-O-H → C-H. Such an operation is needed when an -OH group is replaced by another group, such as an amino group.

Examples

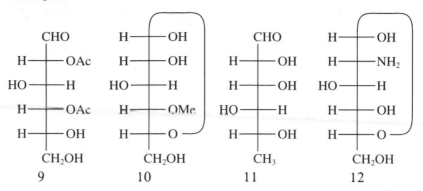

9.   2,4-di-*O*-acetyl-D-glucose (Ac = -OC-CH$_3$)
10. 4-*O*-methyl-α-D-glucopyranose (Me = -CH$_3$)
11. 6-deoxy-D-gulose
12. 2-amino-2-deoxy-α-D-glucopyranose

Acid and alcohol derivatives are named by changing the ending -ose of the saccharide name into the appropriate ending to signify a functional change, for instance, -onic or -aric acid, and -itol.

Examples

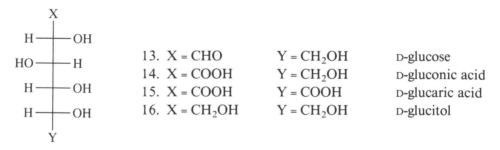

13. X = CHO   Y = CH$_2$OH   D-glucose
14. X = COOH   Y = CH$_2$OH   D-gluconic acid
15. X = COOH   Y = COOH   D-glucaric acid
16. X = CH$_2$OH   Y = CH$_2$OH   D-glucitol

The generic term glycosides defines all mixed acetals formed by the acetalisation of the cyclic forms of aldoses and ketoses. Glycosyl groups are monosaccharides that have lost their anomeric -OH group; the suffix -yl is used to indicate the change that has occurred at C-1.

Examples

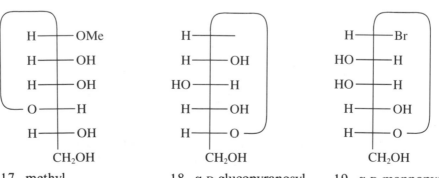

17. methyl
α-D-glucofuranoside

18. α-D-glucopyranosyl

19. α-D-mannopyranosyl
bromide

Disaccharides are named by adding the name of a glycosyl group as a prefix to that of the monosaccharide chosen as parent, as exemplified by α-lactose (a Haworth perspective formula and a conformational formula are shown).

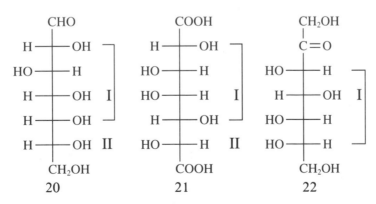

β-D-galactopyranosyl-(1→4)-α-D-glucopyranose

Three forms of abbreviated nomenclature (extended, condensed and short) may be used; D-glucopyranose is represented as D-Glc*p* in the extended form and Glc in the condensed form. The linking atoms are designated by locants, and the α or β configuration of the anomeric carbon atoms is also indicated. The sugar raffinose becomes α-D-Gal*p*-(1→6)-α-D-Glc*p*-(1↔2)-β-D-Fru*f* (extended form) or Gal(α1-6)Glc(α1-2β)Fru*f* (condensed) or Galα-6Glcα-βFru*f* (short).

Aldoses are systematically named as pentoses, hexoses, heptoses, octoses, nonoses, etc., according to the total number of carbon atoms in the chain. The configuration is described by appropriate stereodescriptors (*glycero-* from glyceraldehyde, *gluco-* from glucose, *galacto-* from galactose, etc.) together with the appropriate D or L, and these are assembled according to specific rules in front of the basic name. Names of ketoses are characterised by the ending -ulose.

Examples

20. D-*glycero*-D-*gluco*-heptose (I = D-*gluco*, II = D-*glycero*)
21. L-*glycero*-D-*galacto*-heptaric acid (I = D-*galacto*, II = L-*glycero*)
22. L-*gluco*-hept-2-ulose (I = L-*gluco*)

7.3 NOMENCLATURE AND SYMBOLISM FOR AMINO ACIDS AND PEPTIDES

A peptide is any compound produced by amide formation between a carboxyl group of one amino acid and an amino group of another. The amide bonds in peptides are called peptide bonds. The word peptide is usually applied to compounds whose amide bonds (sometimes called eupeptide bonds) are formed between C-1 of one amino acid and N-2 of another, but it includes compounds with residues linked by other amide bonds (sometimes called isopeptide bonds). Peptides with fewer than about 10–20 residues may also be called oligopeptides; those with more residues are called polypeptides. Polypeptides of specific sequence of more than about 50 residues are usually known as proteins, but authors differ greatly on where they start to apply this term.

Amino acids are represented in two ways: either as $H_2N\text{-}CHR\text{-}COOH$ or as the zwitterionic form $^+H_3N\text{-}CHR\text{-}COO^-$. Although the second of these forms is overwhelmingly predominant in the crystal and in solution, it is generally more convenient to name them and their derivatives from the first form. They are normal organic compounds and are treated as such as far as numbering and naming are concerned, although trivial names are retained for all natural amino acids.

Examples

1. $H_2N\text{-}CH_2\text{-}COOCH_3$	methyl glycinate, glycine methyl ester
2. $CH_3\text{-}CH(NH_2)\text{-}CONH_2$	alaninamide, alanine amide
3. $HOCH_2\text{-}CH(NHCOCH_3)\text{-}COOH$	*N*-acetylserine

There are two generally accepted systems of abbreviation for trivial names, using either one or three letters. The choice of form to use is generally determined by circumstances. Normally, three-letter symbols are used, and one-letter symbols are reserved for long sequences of amino acids. A list of such symbols is shown in Table 7.1.

The stereodescriptors D and L are used with reference to the D configuration of glyceraldehyde.

Table 7.1 Names and their abbreviations and symbols for amino acids.*

Trivial name	Symbols		Trivial name	Symbols	
Alanine	Ala	A	Leucine	Leu	L
Arginine	Arg	R	Lysine	Lys	K
Asparagine	Asn	N	Methionine	Met	M
Aspartic acid	Asp	D	Phenylalanine	Phe	F
Cysteine	Cys	C	Proline	Pro	P
Glutamine	Gln	Q	Serine	Ser	S
Glutamic acid	Glu	E	Threonine	Thr	T
Glycine	Gly	G	Tryptophan	Trp	W
Histidine	His	H	Tyrosine	Tyr	Y
Isoleucine	Ile	I	Valine	Val	V

* The general representation for an unspecified amino acid is Xaa, symbol X.

$$\underset{\text{L}}{\overset{\text{COO}^-}{\underset{\text{R}}{^+\text{H}_3\text{N}\,\text{—}\!\!\!|\!\!\!\text{—}\,\text{H}}}} \quad \underset{\text{L}}{\overset{\text{COO}^-}{\underset{\text{R}}{^+\text{H}_3\text{N}\,\blacktriangleright\text{C}\,\blacktriangleleft\,\text{H}}}} \quad \underset{\text{L}}{\overset{^+\text{H}_3\text{N}}{\underset{\text{COO}^-}{\text{H--C—R}}}} \quad \underset{\text{D}}{\overset{\text{CHO}}{\underset{\text{CH}_2\text{OH}}{\text{H—}\!\!\!|\!\!\!\text{—OH}}}}$$

Three-letter symbols and standard group abbreviations are used to designate amino acids functionalised on -COOH or substituted on $-NH_2$. Modifications are indicated by hyphens in abbreviations.

Examples

   4. *N*-acetylglycine   Ac-Gly
   5. glycine ethyl ester   Gly-OEt
   6. N^2-acetyllysine   Ac-Lys
   7. O^1-ethyl *N*-acetylglutamate   Ac-Glu-OEt

Substitution on other parts of the amino acid is expressed by a different symbolism.

Examples

   8. *S*-ethylcysteine $\underset{\text{Cys}}{\overset{\text{Et}}{|}}$ or Cys(Et)

   9. 3-nitrotyrosine $\underset{^3\text{Tyr}}{\overset{\text{NO}_2}{|}}$

The peptide $H_2N\text{-}CH_2\text{-}CO\text{-}NH\text{-}CH(CH_3)\text{-}COOH$ is named glycylalanine and symbolised as Gly-Ala. The amino acid with the free -COOH group is chosen as the parent. The name of the other amino acid, modified by the suffix -yl, becomes a prefix to it.

The symbolism applied to a peptide is very precise and elaborate. The symbol -Ala stands for $-NH\text{-}CH(CH_3)\text{-}COOH$, and the corresponding name is that of the amino acid. The symbol Gly- means $H_2N\text{-}CH_2\text{-}CO\text{-}$ and corresponds to a name ending in -yl. In the peptide Gly-Gly-Ala, Gly- signifies $H_2N\text{-}CH_2\text{-}CO\text{-}$ and -Gly- -HN-CH_2-CO-, both of these groups being named glycyl, giving the name glycylglycylalanine. A hyphen indicates a C-1-to-N-2 peptide bond. Glutamic acid can be bound through either or both of its two carboxyl groups, and Greek letters α or γ are used to indicate the position of the link.

Example

10. $HOOC\text{-}CH_2\text{-}CH_2\text{-}CH(NH_2)\text{-}CO\text{-}$ α-glutamyl
           5 4   3   2   1
             γ   β   α

The dipeptide *N*-α-glutamylglycine is abbreviated as Glu-Gly. *N*-γ-Glutamylglycine is represented by the bond symbol

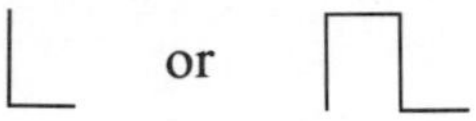

⌐   or   ⌐⌐

Examples

11. 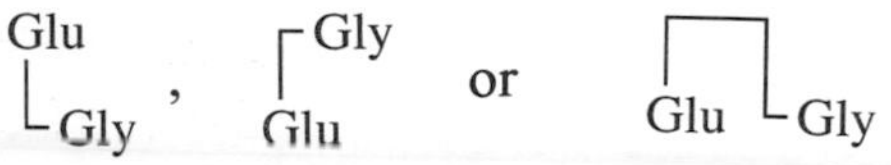 Glu
     |___ Gly ,   ⌐Gly   or   ⌐⌐   or   Glu(-Gly)
                   Glu   Glu └Gly

Glutathione is

12. 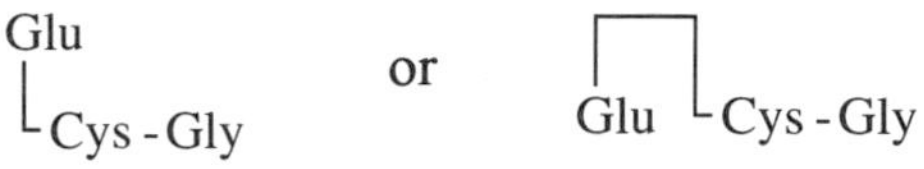 Glu
     |___ Cys-Gly   or   ⌐⌐   or   Glu(-Cys-Gly)
                            Glu └Cys-Gly

Great care should be exercised in presenting these formulae, because a simple line between Glu and Cys would correspond to a thiolester between the γ carboxyl of Glu and the SH group of Cys.

Example

13. 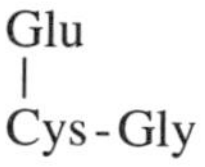 Glu
     |
     Cys-Gly

Cyclic peptides in which the ring consists solely of aminoacid residues with eupeptide links may be called homodetic cyclic peptides. Three representations are possible. Gramicidin S is given as an example. In this decapeptide, all amino acids are L, with the exception of Phe which is D, as is shown by D-Phe or DPhe.

Examples

14. cyclo(-Val-Orn-Leu-D-Phe-Pro-Val-Orn-Leu-D-Phe-Pro-)

15. ⌐Val-Orn-Leu-D-Phe-Pro-Val-Orn-Leu-D-Phe-Pro⌐

16. ⌐→Val→Orn→Leu→D-Phe→Pro⌐
     └Pro←D-Phe←Leu←Orn←Val←┘

Heterodetic cyclic peptides are peptides consisting only of amino acid residues, but the links forming the ring are not solely eupeptide bonds; one or more is an isopeptide, disulfide, ester, etc.

Examples

17. Oxytocin   ⌐⌐Cys-Tyr-Ile-Gln-Asn-Cys-Pro-Leu-Gly-NH$_2$

18. Cyclic ester of threonylglycylglycylglycine   ⌐⌐Thr-Gly-Gly-Gly┘

7.4 LIPID NOMENCLATURE

Lipids are substances of biological origin that are soluble in non-polar solvents. There are saponifiable lipids, such as acylglycerols (fats and oils), waxes and phospholipids, as well as non-saponifiable compounds, principally steroids.

The term 'fatty acid' designates any of the aliphatic carboxylic acids that can be liberated by hydrolysis from naturally occurring fats and oils. 'Higher fatty acids' are those that contain ten or more carbon atoms. Neutral fats are mono-, di- or tri-esters of glycerol with fatty acids, and are therefore termed monoacylglycerol, diacylglycerol and triacylglycerol. Trivial names are retained for fatty acids and their acyl groups: stearic acid, stearoyl; oleic acid, oleoyl. Esters from glycerol are usually named by adding the name of the acyl group to that of glycerol.

Examples

1. 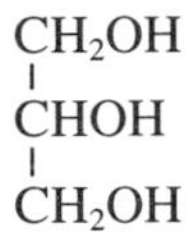
$$CH_2OH$$
$$CHOH$$
$$CH_2OH$$
glycerol

2. 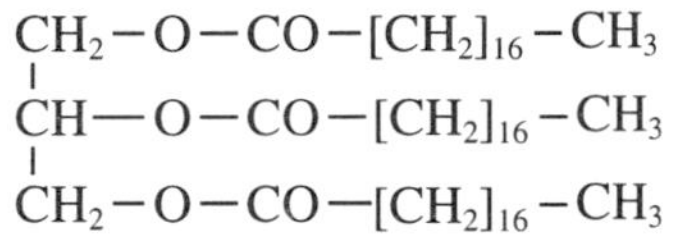
$$CH_2-O-CO-[CH_2]_{16}-CH_3$$
$$CH-O-CO-[CH_2]_{16}-CH_3$$
$$CH_2-O-CO-[CH_2]_{16}-CH_3$$
tristearoylglycerol

3.
$$CH_2OH$$
$$H \qquad NH_2$$
$$H \qquad OH$$
$$CH_2-[CH_2]_{13}-CH_3$$ 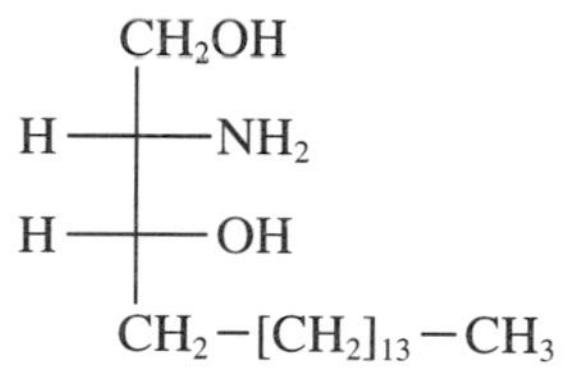
sphinganine

Compounds similar to glycerol, called sphingoids, are derivatives of sphinganine (D-*erythro*-2-aminooctadecane-1,3-diol). The trivial name sphinganine implies the stereochemistry; the use of the stereodescriptor D-*erythro* in the systematic name is to be noted.

Phospholipids are lipids containing phosphoric acid as a mono- or di-ester. When glycerol is esterified on C-1 by a molecule of phosphoric acid, the result is a chiral glycerol phosphate. A specific numbering is necessary to name it without ambiguity. The symbol *sn* (for stereospecifically numbered) is used: thus, D-glycerol and L-glycerol phosphates are easily recognised, the *sn*-glycerol 1-phosphate being the enantiomer of *sn*-glycerol 3-phosphate.

Examples

4. 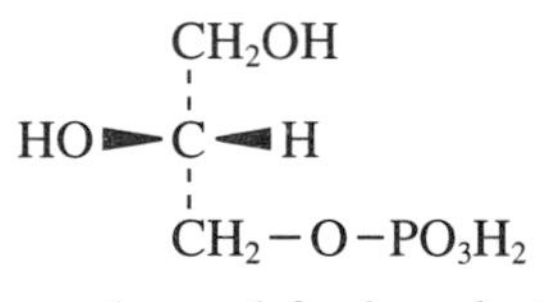
$$CH_2OH$$
$$HO-C-H$$
$$CH_2-O-PO_3H_2$$
sn-glycerol 3-phosphate

5. 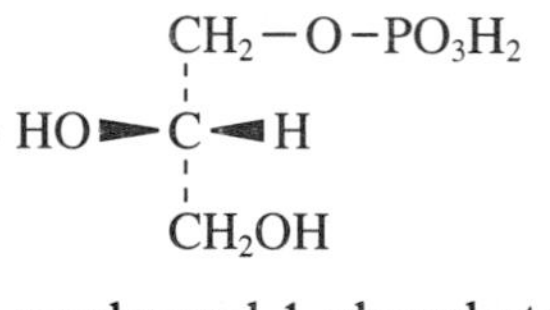
$$CH_2-O-PO_3H_2$$
$$HO-C-H$$
$$CH_2OH$$
sn-glycerol 1-phosphate

A complete name is given for a phospholipid as shown below. Note that the name 'ethanolamine' is an allowed name for 2-aminoethanol.

6. 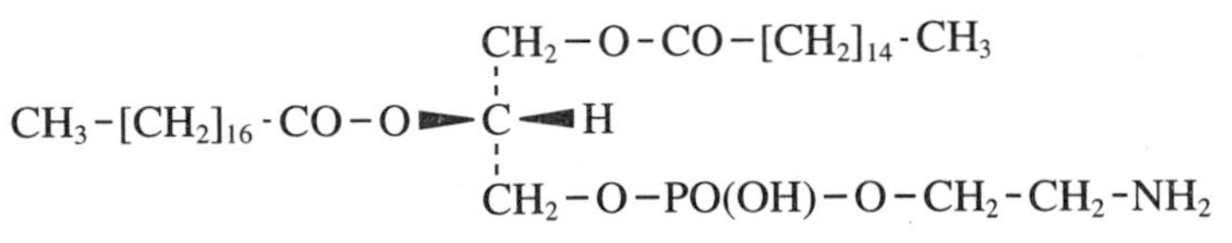
$$CH_2-O-CO-[CH_2]_{14}\cdot CH_3$$
$$CH_3-[CH_2]_{16}\cdot CO-O-C-H$$
$$CH_2-O-PO(OH)-O-CH_2-CH_2-NH_2$$

1-palmitoyl-2-stearoyl-*sn*-glycero-3-phosphoethanolamine

7.5   STEROID NOMENCLATURE

Steroids are compounds possessing the tetracyclic skeleton of cyclopenta[*a*]phenan-threne (1) or a skeleton derived therefrom by one or more bond scissions or ring expansions or contractions. Natural steroids have trivial names. The nomenclature of steroids is not based on these trivial names, but on a few stereoparent hydrides that are common to many compounds. Substitutive nomenclature is used to designate characteristic groups and unsaturation. Structural modifications are expressed by appropriate non-detachable prefixes.

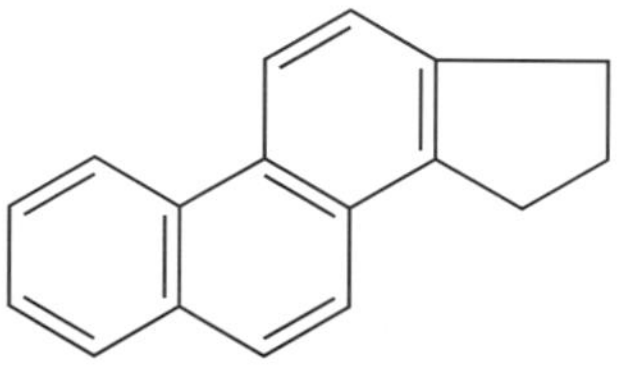

(1) 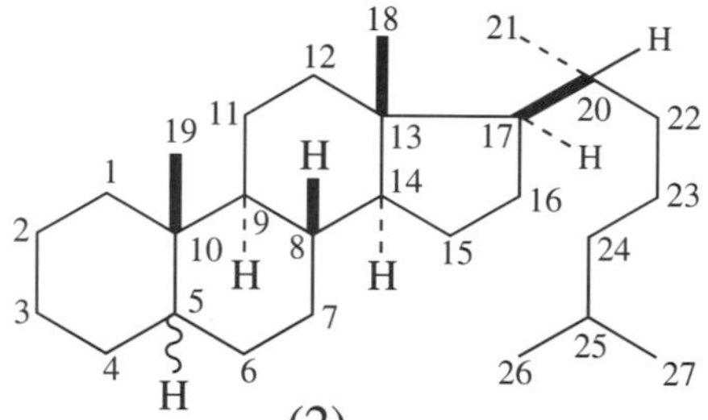   (2)

Some of the stereoparent hydrides are shown below. They represent the absolute configurations and are numbered as shown in (2). According to the projection formula, substituents are designated α (below the plane of the skeleton) and β (above that plane). When no indication is given, the configurations at carbon atoms 8, 9, 10, 13, 14, 17 and 20 are as shown in (2). The configuration at C-5 is either 5α or 5β; it must always be stated in the name. Any unknown configuration is denoted by ξ (Greek xi). In structural formulae broken, thick and wavy bonds (∿) denote the stereodescriptors α, β and ξ, respectively.

Examples

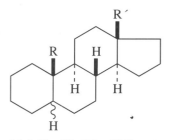

1(a) R = H; R′ = CH₃ estrane
 (b) R = R′ = CH₃ androstane

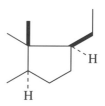

2. pregnane

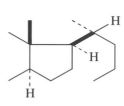

3. cholane

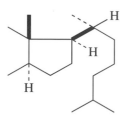

4. cholestane

Examples of trivial and systematic names

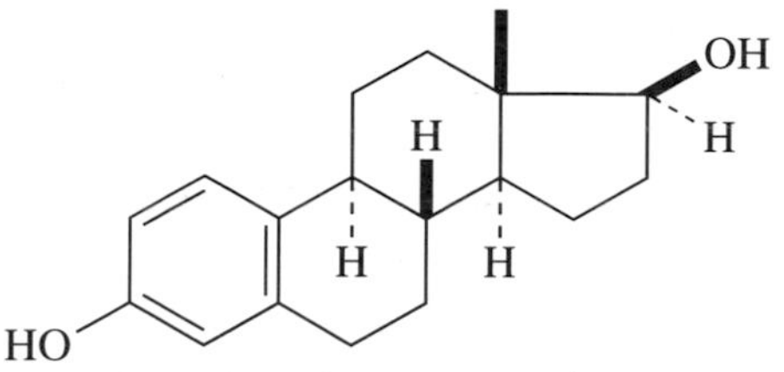

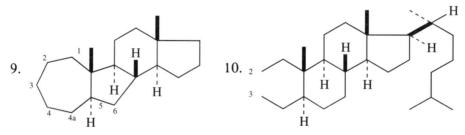

β-estradiol
5. estra-1,3,5(10)-triene-3,17β-diol

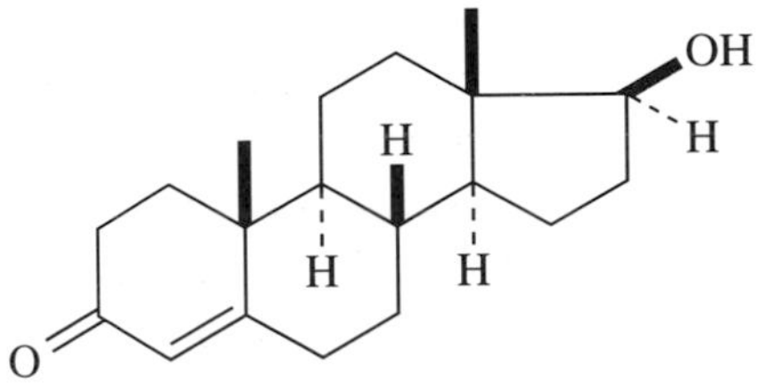

testosterone
6. 17β-hydroxyandrost-4-en-3-one

7.

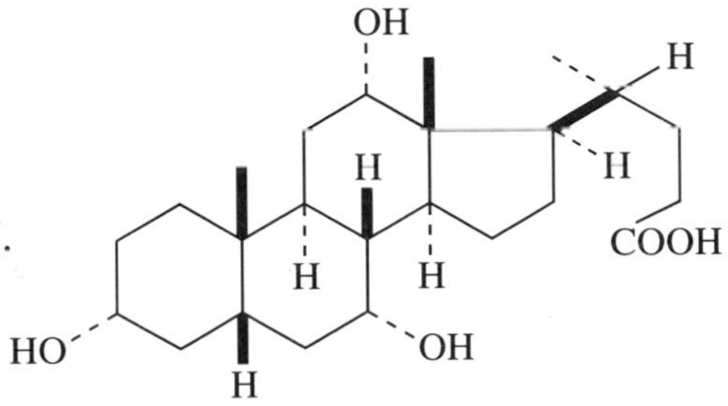

cholic acid
3α,7α,12α-trihydroxy-5β-cholan-24-oic acid

8.

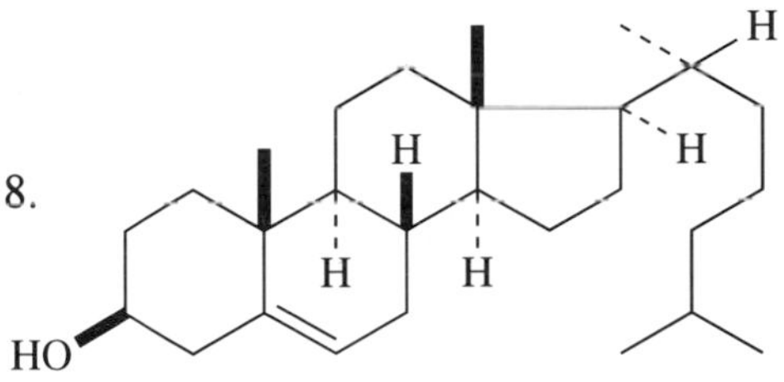

cholesterol
cholest-5-en-3β-ol

Non-detachable prefixes are used to indicate modifications affecting rings: homo-(enlarging), seco- (opening) and nor- (contracting). Locants are supplied as necessary to indicate the positions of modification.

Examples

9.

4a-homo-7-nor-5α-androstane

10.

2,3-seco-5α-androstane

8   Nomenclature in the making

Nomenclature is not a static subject. It changes as new kinds of compound are synthesised and new procedures and devices have to be invented. Fullerenes are a case in point, and their detailed nomenclature and numbering are currently (1997) under discussion. Even established procedures are continuously being reviewed and revised, but the principles established in this survey are likely to remain recommended for the foreseeable future.

As a help in this direction, the CNOC has recently published *A Guide to IUPAC Nomenclature of Organic Chemistry, Recommendations 1993*. It includes revisions, published and hitherto unpublished, to the 1979 Edition of *Nomenclature of Organic Chemistry*. The variations of oxidation state and stereochemistry imposed upon inorganic compounds by the presence of transition metals and non-stoichiometry, etc. make inorganic nomenclature much more difficult to unify. Organic nomenclature is currently better integrated than inorganic nomenclature.

One recent development in substitutive nomenclature that we have employed in this presentation is to place locants immediately before the part of the name (suffix, infix, prefix or descriptor) that they qualify. The CNOC Guide also recommends that the traditional names of alkyl groups be supplemented by alkanyl names that are derived by adding the suffix -yl to the name of the parent hydride, rather than replacing the ending -ane by the suffix -yl, as in current practice. For example, three names will now compete for $(CH_3)_2CH-$: isopropyl, 1-methylethyl and propan-2-yl. The last systematic name is short, easy to understand and to use and fully able to cope with further substitution. The substituted group $(ClCH_2)_2CH-$ can be named 2-chloro-1-(chloromethyl)ethyl or, in the new form, 1,3-dichloropropan-2-yl. This last name is shorter and quite appealing.

The general trend in substitutive nomenclature is to use fewer trivial or retained names and to approach substitutive nomenclature more systematically. A total of 209 trivial names are to be retained: 81 for parent hydrides and 58 (those in Chapter 4, Section 4.5.7, p. 91) for functional parent hydrides. These are so well anchored in nomenclature that they will probably survive for a great many years. Finally, 70 trivial names, such as isobutane and neopentane, are still allowed to be specific for non-substituted compounds. Presumably some of these will be discarded during later revisions of the *Nomenclature of Organic Chemistry*.

The nomenclature of organometallic compounds is currently being reviewed jointly by the CNIC and the CNOC. Some systematic but not widely used names are to be recommended from now on. Names such as phosphine and arsine have been superseded by phosphane and arsane. The names azane and diazane have been suggested to replace ammonia and hydrazine. This is unlikely to happen in the near future, although azo compounds (R-N=N-R) are named by using diazene as a parent hydride. For instance, dimethyldiazene is the substance formerly known as azomethane. Sulfane, H_2S, is used to name sulfides, e.g. diethylsulfane, $(C_2H_5)_2S$, which is also known as diethyl sulfide. The prefix mercapto- is replaced by the systematic name sulfanyl, leading to names such as sulfanylacetic acid for $HS-CH_2-COOH$.

The Commission on the Nomenclature of Inorganic Chemistry is currently producing a further volume of the *Nomenclature of Inorganic Chemistry*, which will deal with more specialised aspects of inorganic nomenclature not currently treated in the 1990 version. For example, one chapter will be devoted to the nomenclature of nitrogen hydrides, another to the nomenclature of iso- and heteropolyanions and yet another to techniques and recommendations for abbreviations of names, especially ligand names. These chapters are innovative but also codify a great deal of established practice.

An area of current development is the nomenclature of organometallic compounds. Organometallic compounds of Main Group elements can, to a first approximation, be considered to be derivatives of hydrides, and the methods of substitutive nomenclature can be applied. Even then, the accessibility of different oxidation states, as with phosphorus(III) and phosphorus(V), introduces complications. Transition metal organometallic compounds are even more difficult to treat, and the development of a unified, self-consistent and accepted and applied nomenclature is not easy. Witness the different ways (κ, η and italicised symbols) for denoting donor atoms in ligands.

In other areas, such as oxo acids, a great deal of traditional, inadequate semi-systematic nomenclature (for example, the names of phosphorus acids) will have to be abandoned before a more rigorous nomenclature can be adopted and generally understood. There is much work yet to do.

The Commission on Macromolecular Nomenclature is currently working on the extension of macromolecular nomenclature to branched and cyclic macromolecules, micronetworks and polymer networks, and to assemblies held together by non-covalent bonds or forces, such as polymer blends, interpenetrating networks and polymer complexes.

The great dream of the founders of systematic organic nomenclature at the Geneva Conference in 1892 was the provision of a unique name for a given compound. This is now becoming feasible, at least in substitutive nomenclature. A unique name is more than ever necessary in legal documents and patents. As trade and commerce become even more international and as general interests, such as the environment, health and safety, become more widespread, the unique name becomes a necessity. Nomenclaturists must reduce choice, and systems and rules should suffer no exceptions. The emergence of computerised nomenclature will help to fulfil that goal.

Index

Page references to tables appear in **bold** type.